ARTIFICIAL INTELLIGENCE FOR SMART TECHNOLOGY IN THE HOSPITALITY AND TOURISM INDUSTRY

ARTIFICIAL INTELLIGENCE FOR SMART TECHNOLOGY IN THE HOSPITALITY AND TOURISM INDUSTRY

Edited by

Vinod Kumar Shukla, PhD

Amit Verma, PhD

Jean Paolo G. Lacap, PhD

AAP | APPLE ACADEMIC PRESS

First edition published 2025

Apple Academic Press Inc.
1265 Goldenrod Circle, NE,
Palm Bay, FL 32905 USA

760 Laurentian Drive, Unit 19,
Burlington, ON L7N 0A4, CANADA

CRC Press
2385 NW Executive Center Drive,
Suite 320, Boca Raton FL 33431

4 Park Square, Milton Park,
Abingdon, Oxon, OX14 4RN UK

Library and Archives Canada Cataloguing in Publication

Title: Artificial intelligence for smart technology in the hospitality and tourism industry/edited by Vinod Kumar Shukla, PhD, Amit Verma, PhD, Jean Paolo G. Lacap, PhD.
Names: Shukla, Vinod Kumar (Professor of information technology), editor. | Verma, Amit (Dean of Student Affairs at Global Business Studies), editor. | Lacap, Jean Paolo G., editor.
Description: First edition. | Includes bibliographical references and index.
Identifiers: Canadiana (print) 20240299760 | Canadiana (ebook) 20240299779 | ISBN 9781774914908 (hardcover) | ISBN 9781774914915 (softcover) | ISBN 9781003432951 (PDF)
Subjects: LCSH: Tourism—Information technology. | LCSH: Hospitality industry—Information technology. | LCSH: Artificial intelligence.
Classification: LCC G156.5.I5 A78 2024 | DDC 910.285—dc23

ISBN: 978-1-77491-490-8 (hbk)
ISBN: 978-1-77491-491-5 (pbk)
ISBN: 978-1-00343-295-1 (ebk)

About the Editors

Vinod Kumar Shukla, PhD
Associate Professor and Head of Academics,
School of Engineering Architecture Interior Design,
Amity University, Dubai, UAE.

Vinod Kumar Shukla, PhD, is currently working with Amity University, Dubai, UAE, as an Associate Professor and Head of Academics in the School of Engineering Architecture Interior Design. He has more than 14 years of experience. He has completed his PhD in the field of "Semantic Web and Ontology." He is an active member of IEEE. He has published many research papers in various reputed journals/conferences. He has also completed the General Management Program from the Indian Institute of Management Ahmedabad (IIM-A). He has conducted many training programs, some of which include a training program for the employees of Delhi Transco Limited (D.T.L), the Indian postal department in Meghdoot Bhawan, Delhi, and Directorate General Resettlement (D.G.R) is an Inter Service Organization functioning directly under the Ministry of Defense (D.G.R). He has also written case studies on "Online Retail in UAE: Driven to Grow" and "Modi's Visit to UAE: Uncovering New Areas of Cooperation," which are hosted by the European Case Clearing House (ECCH), the case center, UK. He has written/edited many books in the field of Industry 4.0 and information technology. He holds international patents in the broader field of information technology.

He received the "Inquisitive Award" in Unanimity (2020), organized and hosted by Amity University Dubai, UAE. He has been awarded the Star Supporter Award (2019), which was presented by the Manzil Center, Sharjah, UAE. Her Highness Sheikha Chaica S. Al Qassimi and Dr. Ayesha Saeed Husaini, Director, Al Manzil Center, Sharjah, UAE, handed over the "Appreciation Certificate and Star Supporter Award (2019)" to Dr. Vinod Kumar Shukla for his continuous support to community engagement and volunteerism. He also received a Memento for Amity University Dubai from Mrs. Gloria Gangte, Deputy Chief of Mission, Embassy of India, Sultanate of Oman, for Guidance Seminar (2016).

Amit Verma, PhD
Dean of Student Affairs, Global Business Studies (GBS) Dubai, UAE

Amit Verma, PhD, has 16 years of experience in the academic and tourism industry, currently holding the position of Dean of Student Affairs at Global Business Studies (GBS) Dubai, United Arab Emirates. He has extensive experience in airlines, hotels, destination management, and online travel companies like Austrian Airlines, Taj Hotel, Crowne Plaza, MakeMyTrip, Travelport, and Carlson Wagonlit Travel. His latest academic qualification is Executive-GMP, a General Management Program from IIM-Ahmedabad.

Moreover, he has been teaching for the last 10 years. He has been associated with Skyline University Sharjah, the Management Development Institute of Singapore, MDIS-Uzbekistan, and Lucknow University. He has been a visiting lecturer for Bangor and Sunderland University in the UK.

Dr. Verma has contributed effectively to the Emiratization plan in Dubai and managed the project of the Ajman Tour Guide Training program. He also initiated the tour guide license framework in the capacity of Head of Department at Skyline University. He has been instrumental in the RAKTA training program and published development initiatives in coordination with the Fujairah Tourism and Antiquities Authority. He has planned and executed the Save Water Program with the Shurooq Authority. He has achieved several awards for successfully operating training programs like the IATA Top Performer Award, ACCA Gold status award, etc.

He has been active in his research interests in the tourism industry and has presented a number of papers focusing on cultural tourism, smart cities, robotics, artificial intelligence, virtual reality, IoT, and other innovation-based solutions in the tourism and hospitality industry.

Jean Paolo G. Lacap, PhD
Vice President, Research and Extension of the City College of Angeles, Angeles City, Philippines

Jean Paolo G. Lacap, PhD, is currently the Vice President for Research and Extension of the City College of Angeles, Angeles City, Philippines. He is a recipient of the 2018 Research Productivity Award bestowed by the Philippine Association of Researchers and Statistical Software Users (PARSSU) because of his outstanding contributions to research publications in SCOPUS and Clarivate Analytics indexed journals. He is also the

2016 Outstanding Business Educator in the field of Entrepreneurship and the 2014 Young Achiever Award in Business Education, both presented by the Philippine Council of Deans and Educators in Business (PCDEB). With his outstanding accomplishments in business education, he was recently conferred the highest recognition of Diplomate in Business Education (DBE) by the Philippine Academy of Professionals in Business Education. He is also a Registered Marketing Professional (RMP) and Accredited Marketing Educator (AMEd).

He has presented 18 research papers at different conferences in the Philippines, Malaysia, Indonesia, Thailand, and Australia and has published 25 research studies in reputable international refereed and SCOPUS and Clarivate Analytics indexed journals, including *Asia-Pacific Social Science Review, Advances in Hospitality and Tourism Research, Journal of Science and Technology Policy Management*, to name a few. He has received several research awards and recognition as Best Research Paper and Best Research Paper Presenter. He is also part of several international refereed journals in the Philippines and abroad as an editorial board member and reviewer. He is also a member of the International Steering Committee of the Technology Innovation Management and Engineering Science International Conference, a conference being organized by Mahidol University, the leading university in Thailand.

Contents

Contributors

Abhishek Achary
Department of Humanities, Arts and Applied Sciences Amity University, Dubai, UAE

Diogo Davidson Albuquerque
Department of Humanities, Arts and Applied Sciences Amity University, Dubai, UAE

Saniya Ansari
Department of Engineering and Architecture, Amity University, Dubai, UAE

Fatima Beena
De Montfort University, Dubai, United Arab Emirates

Prashant Chauhan
Research Scholar, School of Law, University of Petroleum and Energy Studies, Dehradun, Uttarakhand, India

Soumi Dutta
Institute of Engineering and Management (IEM), West Bengal, India

Muhammet Erbay
Faculty of Tourism, Selcuk University, Turkey

Shakhnozakhon Fakhriddinova
Specialist of the Research and Analytical Activities, Department of the Tourism Innovation and Co-Learning Center, Management Development Institute of Singapore in Tashkent, Tashkent

Wonda Grobbelaar
Nature Science Research and Innovation Center (NSRIC), Canada

Shaurya Gupta
School of Computer Science, University of Petroleum and Energy Studies (UPES), Dehradun, Uttarakhand, India

E. Eze Ifeanyichukwu
Diversity and Inclusion Specialist, Royal Springs International Abuja, Nigeria

Vineeta Kapoor
Freelancer Writer, Alumnus (2006) – Amity University, Noida, Uttar Pradesh, India

Venkat Sai Karanam
Department of Engineering and Architecture, Amity University, Dubai, UAE

Gagandeep Kaur
School of Law, University of Petroleum and Energy Studies, Dehradun, Uttarakhand, India

Mohammad Nadeem Khalid
Institute of Applied Technology, Abu Dhabi, UAE

Cari Leyshon
CEO and Founder, CQify, Spain

Darryl Mathew
Department of Engineering and Architecture, Amity University, Dubai, UAE

Sana Moid
Amity Business School, Amity University, Lucknow Campus, Uttar Pradesh, India

Ashish Raina
Kanya Maha Vidyalaya, Jalandhar, Punjab, India

Varinder Singh Rana
Department of Hospitality and Tourism Management, City University College of Ajman, UAE

Nitin Shankar
Amity Business School, Amity University, Lucknow Campus, Uttar Pradesh, India

Ankita Sharma
Student, Global Business School, Dubai, UAE

Robin Sharma
Department of Humanities, Arts and Applied Sciences Amity University, Dubai, UAE

Vinod Kumar Shukla
Department of Engineering and Architecture, Amity University, Dubai, UAE

Atul Abraham Thomas
Department of Humanities, Arts and Applied Sciences Amity University, Dubai, UAE

Amit Verma
HTMi Switzerland, Dubai, United Arab Emirates

Sonali Vyas
School of Computer Science, University of Petroleum and Energy Studies (UPES), Dehradun, Uttarakhand, India

Abbreviations

AI	artificial intelligence
AIaaS	artificial intelligence as a service
AR	action research
AR	augmented reality
AWS	Amazon web services
BI	business intelligence
CQ	cultural intelligence
CRM	customer relationship management
CRO	chief robotic officers
CRSs	central reservation systems
CV	curriculum vitae
DaaS	diversity as a service
EVI	enterprise visual intelligence
F&B	food and beverage
GDP	gross domestic product
GDPR	general data protection regulation
GDSs	global distribution systems
GPS	global position systems
HACCP	hazard analysis critical control points
HDFS	Hadoop distributed file system
HDM	high-definition mobile
HRI	human-robotic interaction
IaaS	infrastructure as a service
IDC	International Data Corporation
IIoT	industrial internet of things
IoT	internet of things
IT	information technology
KYC	know-your-customer
MICE	meetings, incentives, conferences, and exhibitions
MIT	Massachusetts Institute of Technology

MR	mixed reality
MSR	Microsoft Research Asia
NLP	natural language processing
OTP	one-time password
PDAs	personal digital assistants
PIR	passive infrared sensor
PMS	property management system
PUCL	people's union for civil liberties
QR	quick response
RAISA	Robotics, Artificial Intelligence, and Service Automation Association
RPA	robotic process automation
SaaS	software as a service
SERP	search engine results page
STR	Smith travel research
UAE	United Arab Emirates
UGC	user-generated content
UID	unique identifier
VR	virtual reality
WTTC	World Travel and Tourism Council

About the Book

This book is focused on the shifting requirements of the service industry, leading to multidimensional changes in the pivoting industry scenarios along with the complex situations that transformed the teaching, learning, and operation of each service or non-service-related industry. The book is designed to provide the factors, attributes, and cases that may align learners close to the progressing industry and deliver theories to plan their success strategies based on explicit real-life scenarios.

The book has been designed considering information technology as a prime focus that has played an unprecedented role in the evolution of the service industry from time to time. The authors could not find any remarkable work of such a combination amongst the interdisciplinary books. This book is a fantastic collection of the latest technologies and their applications in the service industry, with real-life cases to disclose the strategic management of technology in the industry.

The book includes major technologies utilized in the service industry: the internet of technology, artificial intelligence, robotics, big data analysis, virtual reality (VR), and augmented reality. Industries like airlines, hospitality, banking, transportation, healthcare, entertainment, travel, food and beverage (F&B), wellness, and culture can be included in the context of the people for their convenience and comfort along with their business requirements.

The content of the book is divided into three different sections. Section I: Artificial Intelligence and Smart Hospitality, is a collection of chapters related to artificial intelligence involvement in smart hospitality. Section II: Artificial Intelligence Enabling Smart Tourism, is the collection of chapters related to artificial intelligence and smart tourism. Section III: Artificial Intelligence Empowering Service Industry, is based on the broader aspect of artificial intelligence in the service industry, with the main focus on customer service, individual privacy, and cybersecurity.

The excellent evolution of the industry and the technological impacts have revolved around the industry point of contact and transformed the business entirely. Henceforth, pivoting industries require students and learners with the latest pool of knowledge to plan and approach the new trends more scientifically.

The book 'Artificial Intelligence for Smart Tourism and Hospitality' is a reference book written specifically for service industry professionals with a focused approach towards introduction to Industry 4.0-related technologies. This book will act as a guide for Industry 4.0 technologies in the service industry. However, it will explain how Artificial Intelligence will support your company's strategy to revolutionize the business by imparting the know-how to use it most effectively.

The book is research-oriented, providing a common platform for all researchers in this domain. It covers different verticals of industry and academics, which is an added advantage of this book. This book would provide a path to walk through the "must-know" and "need-to-know" aspects of artificial intelligence, natural programming language processing, and speech recognition. the versatile conceptualized approach and dynamic industry examples of machine learning depict the optimum utilization of the right technology in the vast scope of the service industry.

Preface

This book delves deep into the ever-evolving demands of the smart tourism and hospitality industry. It explores the multitude of changes that have occurred in the industry, resulting in complex situations that have completely transformed the way services are taught, learned, and operated. This comprehensive book aims to equip learners with the necessary knowledge and understanding to navigate the dynamic landscape of the hospitality and tourism industry. By providing a wealth of factors, attributes, and real-life cases, it empowers learners to develop effective strategies that align with the industry's progress and ensure their own success.

The book, meticulously designed with a keen focus on artificial intelligence, delves into the profound impact this cutting-edge technology has had on the ever-evolving hospitality and tourism industry. Remarkably, the authors have discovered a dearth of literature that explores the interdisciplinary nature of this subject matter. This book, therefore, stands as a remarkable collection of the latest technological advancements and their practical applications within the realms of hospitality, tourism, and service industries. By presenting real-life cases, it unveils the strategic management of technology in these sectors.

Within its pages, one will find an exploration of the major technologies that have revolutionized the service industry, including the Internet of Things (IoT), artificial intelligence, robotics, big data analysis, virtual reality, and augmented reality. Moreover, the book takes into account a wide range of industries, such as airlines, hospitality, transportation, healthcare, entertainment, travel, food and beverage, wellness, and culture, catering to the needs and desires of individuals while also addressing their business requirements. Additionally, the book delves into the crucial aspects of cyber security and privacy, ensuring that readers are well-informed about the protection of their personal information within these industries.

The remarkable evolution of the industry, driven by technological advancements, has greatly influenced the way businesses operate and interact with their customers. This transformation has necessitated the industry to seek out individuals who possess the most up-to-date knowledge and skills in order to effectively navigate and capitalize on the latest trends. It is with this objective in mind that the book "Artificial Intelligence for Smart Tourism and Hospitality" has been meticulously crafted.

This reference book is specifically tailored for professionals working in the service industry, offering a focused approach to introducing the various technologies related to artificial intelligence. Not only does this book serve as a comprehensive guide to understanding and implementing artificial intelligence in the hospitality, tourism, and service sectors, but it also delves into how this cutting-edge technology can be strategically leveraged to revolutionize businesses. By providing invaluable insights and practical know-how, this book equips companies with the necessary tools to harness the power of artificial intelligence and drive their organizations towards unprecedented success.

The content has been designed in three sections with the perception of artificial intelligence and related technology. These three sections focus on smart hospitality, smart tourism, and the service industry. The book is research-oriented, which will provide a common platform for all researchers in this domain. It covers different verticals of industry and academics, which is an added advantage of this book.

CHAPTER 1

Exploring Artificial Intelligence in Hotels

ATUL ABRAHAM THOMAS and DIOGO DAVIDSON ALBUQUERQUE

Department of Humanities, Arts and Applied Sciences Amity University, Dubai, UAE

ABSTRACT

The Hotel Industry is booming with the increase of tourism around the world. With many hotels opening up, there will be an increase in competition and pressure to maintain their standard performance and keep up with the current trends. Automation and machine industries have been integrating new technology and revolutions for business development with digital technological aspects in recent decades. The hotel industry uses various innovative methods to provide specialized customer service and advance. The hotel industry, which has adopted many innovative methods for providing satisfying customer service, has advanced its entire system with the adoption of many comfort-defining advancements. The hospitality industry is enhancing and adapting new modern technology and digital aspects in hotel operations. This chapter explains the importance and disadvantages of AI in the hotel industry and also focuses on why AI can revolutionize the whole industry. The chapter also explains about the concepts of biasness, which may happen due to artificial intelligence (AI).

1.1 INTRODUCTION

Artificial intelligence (AI) is a new branch of the scientific field that has contributed vastly to our daily lives and has become an indispensable tool.

Artificial Intelligence for Smart Technology in the Hospitality and Tourism Industry.
Vinod Kumar Shukla, Amit Verma, and Jean Paolo G. Lacap (Eds.)

It has grown at a rapid rate in the 21st century. AI has evolved from humble beginnings to a worldwide phenomenon. AI is a tool that enables people to reconsider how they combine information, interpret it, and use the consequent insights to make better choices; it is already reshaping every aspect of human existence [1].

AI rapidly improves numerous human tasks, including disease diagnosis, language translation, and customer service. AI can be seen in almost every sector. The hotel industry is undergoing a significant transformation because of AI technological developments in service tasks. Technology exists in almost every sector of tourism and travel. Some of the possible uses could be customization and recommendation systems, robots, dialogic systems, smart travel agents, forecasting, and predicting systems, language translators, voice recognition and natural language processing (NLP) systems [2].

1.2 TOURISM AND HOSPITALITY INDUSTRY: AI PERSPECTIVE

What exactly is AI? Let us travel back to 1998, when John McCarthy, one of the first founders of AI, posed the question. He says, "It is the art and science of creating intelligent machines, brilliant computer software. It is similar to the task of using computers to understand human intelligence, but AI does not have to be limited to biologically observable methods" [1].

AI is being used significantly by modern technology to establish essential business tasks. AI handles a variety of traditional hotel operational tasks. Paperless check-in, guest information collection, billings, and payment modes via electronic methods enhanced by computer technology show that AI would be able to handle various personal tasks as well as customer connection and service [2]. The hospitality industry is revolutionizing at an incredible rate, from pre-customer service interactions to post-customer arrival. AI greatly impacts business acceleration, allowing hotel systems to perform beyond their wildest dreams [3] (Figure 1.1).

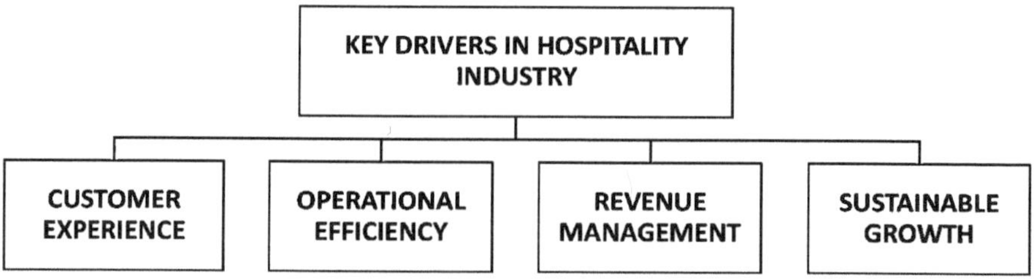

FIGURE 1.1 Key drivers in hospitality industry.

With a rise in travelers, service providers must highlight themselves by providing highly personalized services while managing to keep guest preferences in mind. Understanding customers' preferences and monetizing guest records data analysis is complex. Machine learning is the ideal medium for learning and acting based on previous experiences and enhancing organizational efficiency to meet client choices cost-effectively. A deep understanding of individual consumer tastes allows the customer service representative or marketing team to recommend products and services that the customer is likely to use.

1.3 ARTIFICIAL INTELLIGENCE (AI) SYSTEMS AND THEIR USES

1. **Service Robots:** Mortimer was the world's first service robot designed to serve hotels. The German robot can assist hotel staff in porter services, serving breakfast, and distributing mail to guest rooms. There were plenty of challenges that had to be resolved while developing Mortimer. First and foremost, the robot's size needed to be small to fit through tight spaces while also being large enough to support the weight of the customer's luggage. Moreover, the service robot had to have an accurate sensor system to move fast and avoid bumping into customers. Finally, the service robot needed to be familiar with the hotel's layout, including elevators, bells, and automatic doors. These robots are programmed to perform specific tasks after being fed all necessary information such as layout, service points, holding equipment, and equipment delivery [4].
2. **Chatbot:** Also known as a chatterbot, is software that can communicate with humans using natural conversational language [5, 6]. Companies can use two types of chatbots for customer service: task-oriented chatbots that focus on performing one function based on pre-programmed "rules" and data/AI-driven chatbots that are conversational and self-learning [7].

The advantages of chatbots by which hospitality organizations can benefit are:

- Improves guest satisfaction and loyalty, thus enhancing guest experiences.

- Front-desk staff can devote more time to building personal relationships with guests as the chatbots do the time-consuming work.
- Respond to inquiries more quickly and effectively.
- Ensure more accurate service, lowering costs and eliminating human error risk.
- Improved customer relations for business teams on a larger scale.
- With more personalized engagement, one can identify unique upselling and cross-selling opportunities.

Chatbots can also be used in the hotel, both virtual and in-person. People from all around the globe are already accustomed to texting online and participating in online forums. As a result, the hotels have made a natural progression to AI chatbots. Chatbots save human staff time by assisting with check-in and recommending hidden gems only locals know about [4].

1.4 BENEFITS OF ARTIFICIAL INTELLIGENCE (AI) IN HOTELS

1. **Guest Segmentation and Targeting That is Data-Driven, Automated, and Granular:** Hotels' internal services and customer support must be cost-efficient, and operating charges must be minimized. By combining AI and IoT in the hospitality industry, ML can offer tailor-made solutions. With the help of IoT, sensor systems, and ML, hotels can spot trends, create customized offers, and provide personalized recommendations that complement guests' lifestyle choices and elevate their experience [2].
2. **Enhances Brand Loyalty:** The hospitality industry thrives on its good name. Hotels can now use Deep Learning to analyze online feedback from customers across social media sites and other review forums. They include information about how customers feel about a brand. Positive feedback will improve the establishment's image, while negative feedback will give them a chance to serve the guests better [8].
3. **In-Person Customer Service:** Room bookings, reservations, check-outs, customer service, fundamental problem solving, and even assisting with hotel services are all examples of typical tasks that can be completed without an employee's use [7]. Connie, an

AI robot developed by the Hilton group, is the most outstanding example of this yet. Customers can interact with the robot to receive tourist information. It is imposing that it can understand the speech and adjust to new people—subsequently, the more customers who communicate with it, the better.

4. **Reduce Operation Costs with Revenue Management:** AI is now being used to prompt and manage hotel operations, lowering operating costs while also affecting hotel employment. In comparison to humans, AI provides perfection by reducing operational waste. AI technology reduces operational costs by streamlining processes across the board. The property management system (PMS) satisfies organizational requirements while also providing low-maintenance benefits [7].
5. **Robots in Hotel Industry:** Robotics is an exciting and common trend in the hospitality industry. The Henn-na Hotel, situated in Nagasaki, Japan, was the world's 1st hotel entirely employed by robots. Robots are stationed throughout the hotel to assist with simple tasks such as serving breakfast, basic cleaning, delivering newspapers, carrying luggage, and some food and beverage (F&B) service in restaurants [4, 7].

1.5 CHALLENGES OF ARTIFICIAL INTELLIGENCE (AI) IN HOTELS

The difference between both hospitality and service is thin. Being hospitable is the ability to make others feel valued and welcomed. Service denotes what one does for someone, such as offering breakfast; hospitality denotes how one makes the person feel after the breakfast is offered. Hotel staff must express compassion for the well-being and quality of the visitor's stay to be accepted and hospitable. The combined effect of hospitality and service impacts the customer's experience. A splendid service alone is never recognized alone; hospitality is what brings an emotional attachment. A simple smile could indeed make a significant change. Robots are designed to carry out service tasks such as checking guests in and out, delivering room service, and responding to queries. However, due to their lack of emotion, they will never interact hospitably with guests.

A warm and inviting mindset is anticipated as a hospitable behavior. Respect, openness, humility, curiosity, presence, nonjudgment, and dialog are all elements of a welcoming mindset. Everything in the hotel industry revolves around the ability to read and respond to emotions. Robots fail to meet warm approaches. They are just computer programs incapable of making emotional attachments on a human level. Hotel robots could jeopardize the essential nature of hospitality, service security, and efficiency. This is an unintended consequence for the guests and hotels.

- The first and most pressing issue with AI is how tourists perceive and think about these technologies.
- Data privacy is one of the major challenges that all businesses seeking to leverage AI will face. Data obtained by the AIs during customer interactions may be used for other purposes, posing a violation of privacy.
- Technical malfunctioning cannot be completely avoided. Guests may receive false and misleading information, resulting in mistakes and negative experiences.
- Although access to this technology is improving, it is still a very costly and complex procedure for a company to consider.

Despite this, the absence of human interaction outweighs all of the advantages that robots can provide to hotels., which hurts the guest's overall experience. Due to robots' inability to demonstrate expected hospitable behavior, check-in/check-out, porter features and robotic room service are all undervalued and merely viewed as service components. Hospitality and service are inextricably linked, and the absence of one would tarnish the guest's experience [7].

1.6 ETHICS AND BIASES

Even though technology has come a long way, the field of AI is still in its early stages. AI is expected to impact every aspect of people and culture significantly. It is similar to machines and computers as part of the 4th Industrial Revolution. This raises several ethical issues that must be addressed [9]. Another significant and pertinent risk that comes with the pervasive usage of AI is bias. All human beings are biased, and bias is natural in humans, such as bias based on race, age, sex, and financial standing. The issue is that AI is far more powerful than humans, potentially allowing biases embedded in algorithms to be amplified. As AI actions are based on human input, it is possible to feed it subjective or incorrect data, resulting in decisions that are biased, inappropriate, or irrelevant. This could imply that as AI develops, it will learn to enforce biased structures that will then be replicated. It has been suggested that AI systems be transparent, robust enough to withstand manipulation, and predictable.

While this disadvantage may be overcome as technology advances, it is clear that AI cannot be trusted to make critical decisions at this time. AI systems will be forced to make trade-offs regularly, and they must be able to make fair decisions that will benefit all parties. Finally, if superintelligence systems are developed, ethical considerations should be built into the AI systems [8]. Technology is playing a very vital role in embracing the experience of customers in the hospitality industry. Each of these forces is discussed in detail in subsections. Internet of Things (IoT) and AI have redefined the way today's tourism industry operates [10, 11]. Hospitality industry also got the same benefits and challenges from the technology as tourism industry. Human-robotic interactions (HRIs) are growing nowadays and are also preferred by customers [12]. Technology like blockchain can be adapted in food trackability [13]. Even technology is helping in managing and making customer aware about the halal food [14]. From a sustainability point of view also, technology is helping to manage the entire supply chain in the more effective way [15].

1.7 FUTURE OF ARTIFICIAL INTELLIGENCE (AI)

The future of AI technology in the hotel industry is still uncertain. On the one hand, the perspective is reassuring. As per this point of view, society can tackle AI's most pressing issues. Since privacy concerns will be settled, connectivity will be implemented to allow bots to be implemented, and laborers and bots will function well together. Here, under the framework, AI can be represented as a set of innovations that enhances the tourist stay. Enterprises would get to analyze their guests and thus develop products, services, and experiences more customized to their needs. Businesses could then create flexibly personalized packages centered on their customers' preferences. On the other hand, some people are more pessimistic [9]. According to this point of view, machines will eventually replace a huge percentage of jobs in the industry, giving rise to the extinction of the hotel experience. The usage of machines would be mandatory rather than an option, as it is in many airports already. Guests would then come across devices and robotic systems that are not yet ready for use in a production line, and businesses will choose lower-cost products, even if they perform poorly. In most cases, technology will replace humans, but the potential labor costs would not be carried on to consumers, who would still pay the same but have a poorer experience.

1.8 CONCLUSION

AI has a bright prospect in this hotel industry, from helping concierge staff answer customers' questions to offering personalized suggestions via specific browsers. While AI is still in its infancy, it is already impacting the guest experience throughout the travel lifecycle, including on-property. By delivering sumptuous tech-savvy environments, maximizing efficiency and productivity, and searching for new revenue sources, AI alleviates the pressure of providing an exceptional experience. To meet rising demands and stay ahead of the competition, AI-based innovation is necessary to upgrade services and improve customer experience. However, some significant challenges must be overcome before the technology becomes the new best friend for hoteliers and visitors. The limitations may seem to be barriers, but given the AI solutions that tourism and hotel management can bring, it is critical to overcome them and use AI to see significant changes in operations. If these conditions are met, it is safe to say that AI will undoubtedly transform the hotel industry. The transformation will only accelerate in the coming years [9].

KEYWORDS

- **artificial intelligence**
- **automation**
- **hospitality industry**
- **hotel industry**
- **innovation**
- **tourism**

REFERENCES

1. Bartneck, C., Lütge, C., Wagner, A., Welsh, S. (2021). What Is AI?. In: *An Introduction to Ethics in Robotics and AI*. Springer Briefs in Ethics. Springer, Cham. https://doi.org/10.1007/978-3-030-51110-42.
2. Liu, J., Kong, X., Xia, F., Bai, X., Wang, L., Qing, Q., & Lee, I. (2018). Artificial intelligence in the 21st century. *Ieee Access, 6*, 34403-34421.
3. Al-shami, S. S., Al Mamun, A., Musa, A. E., & Rashid, N., (2021). *Artificial Intelligent Towards Hotels' Competitive Advantage*. An exploratory study from the UAE.
4. Mieczyslaw, L. O. P. W., (2021). *A Note on the Applications of Artificial Intelligence in the Hospitality Industry: Preliminary Results of a Survey*. Elsevier B.V.
5. Suhel, S. F., Shukla, V. K., Vyas, S., & Mishra, V. P., (2020). Conversation to automation in banking through chatbot using artificial machine intelligence language. In: *2020 8th International Conference on Reliability, Infocom Technologies and Optimization (Trends and Future Directions) (ICRITO)* (pp. 611–618). doi: 10.1109/ICRITO48877.2020.9197825.
6. Athota, L., Shukla, V. K., Pandey, N., & Rana, A., (2020). Chatbot for healthcare system using artificial intelligence. In: *2020 8th International Conference on Reliability, Infocom Technologies and Optimization (Trends and Future Directions) (ICRITO)* (pp. 619–622). doi: 10.1109/ICRITO48877.2020.9197833.
7. Yadav, S. (2020). AI in the hotel industry: Adoption and application of artificial intelligence in hotel industry. *Hospitality & Travel Connaisseur Magazine*, 30–32.
8. *Qualetics AI Management System (AIMS)*, (n.d.). Retrieved from Qualetics Data Machine Web Site: https://qualetics.com/ai-in-hospitality-benefits-applications-usecases/ (accessed on 07 September 2023).
9. Bulchand-Gidumal, J., (2020). *Impact of Artificial Intelligence in Travel, Tourism, and Hospitality*. Springer.
10. Verma, A., Shukla, V. K., & Sharma, R., (2021). Convergence of IoT in tourism industry: A pragmatic analysis. In: *Journal of Physics: Conference Series* (Vol. 1714, No. 1, p. 012037). IOP Publishing.
11. Verma, A., & Shukla, V., (2019). Analyzing the influence of IoT in tourism industry. In: *Proceedings of International Conference on Sustainable Computing in Science, Technology and Management (SUSCOM)*. Amity University Rajasthan, Jaipur-India.
12. Grobbelaar, W., Verma, A., & Shukla, V. K., (2021). Analyzing human-robotic interaction in the food industry. In: *Journal of Physics: Conference Series* (Vol. 1714, No. 1, p. 012032). IOP Publishing.
13. Chopra, A., & Shukla, V. K., (2022). Application of disruptive technology in food trackability. In: *ICT and Data Sciences* (pp. 47–62). CRC Press.
14. Petiwala, F. F., Nawazish, H. N., Shukla, V. K., Sharma, R., & Nanda, I., (2021). Halal food benefits and challenges: A mobile application based technical review. In: *2021 International Conference on Advance Computing and Innovative Technologies in Engineering (ICACITE)* (pp. 139–144). IEEE.
15. Albuquerque, D. D., Shukla, V. K., Verma, A., Tyagi, S. K., & Sharma, P., (2020). Enhancing sustainable customer dining experience through QR code and geo-fencing. In: *2020 International Conference on Computation, Automation and Knowledge Management (ICCAKM)* (pp. 190–196). doi: 10.1109/ICCAKM46823.2020.9051470.

CHAPTER 2

Big Data Analytics and Smart Hotels

SANIYA ANSARI,[1] VINOD KUMAR SHUKLA,[1] SONALI VYAS,[2] and SHAURYA GUPTA[2]

[1]*Department of Engineering and Architecture, Amity University, Dubai, UAE*

[2]*School of Computer Science, UPES, Dehradun, Uttarakhand, India*

ABSTRACT

The main aim of this chapter is to find a way for a better Hotel Management System that would address the requirement of customers and would not take a toll on the mechanical network of the system. This chapter focuses on providing solutions to the limitations in the hospitality industry, such as lack of consistency in cleanliness management, difficulties in managing energy resources, and addressing the security and safety challenges in the hotel environment. The chapter uses a conceptual approach where the current information about IoT in the hospitality industry is considered to calculate limitations and provide corresponding solutions. The chapter uses secondary data and information from other sources such as journals, conference papers, research articles and the Internet. This chapter uses Big Data analytics using the 'divide and conquer' methodology to create a framework of the hotel system to find solutions for limitations using the Internet, cloud, devices or things, sensors, smart devices, image recognition and data analysis as they appear. It also aims to provide methods to make the solution more practical and spread it to hospitality industry segments like hotels, resorts, and many more. The chapter also discusses the survey result, which focuses on understanding the preference of technology over services. This system could change the mechanism of standard hotels as

Artificial Intelligence for Smart Technology in the Hospitality and Tourism Industry.
Vinod Kumar Shukla, Amit Verma, and Jean Paolo G. Lacap (Eds.)

people prefer these services over regular hotels. It could bring the change we wish to see in the hospitality industry.

2.1 INTRODUCTION

The IoT refers to the millions of peripheral devices connected by the internet. They all receive and share data because of the prevalent use of very cost-effective computer chips and the growing field for the facilities of wireless networks. It can turn any dumb machine into a smart device. Something as small as a keyhole can be connected to something as big as a ship. Connecting these devices has enabled the speed of the network to increase, thereby increasing the sending of data. Adding sensors adds a new level of intelligence to devices that were rather dumped, enabling them to function without human intervention. Real-time data can be shared and received with the help of the sensors. The IoT is building smarter webwork worldwide by enabling machines to think and transmit data, merging digital and real-world. The hospitality industry includes luxury leisure locations like hotels, resorts, and tourism destinations, which serves as a relaxation time for people worldwide. The lodging business is reliably developing, with more than 700,000 inns throughout the planet and offering more than $3.41 trillion to the worldwide economy; the business has recuperated well from the 2008 downturn. In 2019, the worldwide lodging industry was valued at USD 570 billion. Beijing, China, is home to the most inns on the planet, with more than 4,169. In the US, the city with the most lodgings in Las Vegas, with more than 150,000 rooms, and New York comes in second with more than 115,000. With the developing interest for movement, the business is proceeding to develop quickly; the number of new inns opening simply in London will reach 121, offering 14,840 new rooms in 2019–2020. Combining these two vastly expanding fields could create a solution that significantly benefits both sides [1].

2.2 HOSPITALITY INDUSTRY DURING COVID-19

Fundamental discoveries also show that many hotel clients and most of the inn clients accept that using different advancements in assistance conveyance will be essential in the COVID-19 climate to limit human-to-human contact. Late improvements in artificial brainpower and social help robot advances have empowered the utilization of AI innovations in assistance conveyance. The COVID-19 Pandemic may hasten the ubiquity of such innovation for public security reasons. Subsequently, it is pivotal for accommodation analysts to explore what AI gadgets used in help conveyance will mean for activities, workers, and clients. Moreover, it is essential to recognize the physical and mental components that can impact clients' and representatives' acknowledgment of AI gadget use in help conveyance.

The Pandemic's impact has severely impacted the economy of the countries dependent on tourism. The Tourism industry, responsible for 10% of GDP (gross domestic product) and more than 320 million jobs, have been impacted with an intense fallback of 100 jobs at risk and would take its time to recover, expected to be 2023. The Pandemic was announced at the start of the year. With its influence, half of the countries adopted strict lockdown of the borders with immediate effect in February resulted in a 65% fall in tourist arrival [2]. The occupancy fell by 44% and the room rates by 21%, as per the STR (Smith Travel Research) report. This has forced hoteliers to be more creative and start planning strategies for new markets and exploring new clients. Hoteliers also focused on diversifying their F&B services to turn them into profit centers [3].

The WTTC (World Travel and Tourism Council) has stated it correctly that the patches of policies and practices run by all the countries to deal with the Pandemic will not be satisfying nor complement the full-fledged movement. However, implementing the right technologies in the processes would enable the countries to smoothly welcome the tourists and guests by sharing relevant data through digital travel portals and apps [4]. UAE (United Arab Emirates), one of the fast-paced counties, has recovered and organizing the Expo 2020 considered in the MICE (meetings, incentives, conferences, and exhibitions) industry has resulted in a decline in tourist arrivals (Figure 2.1).

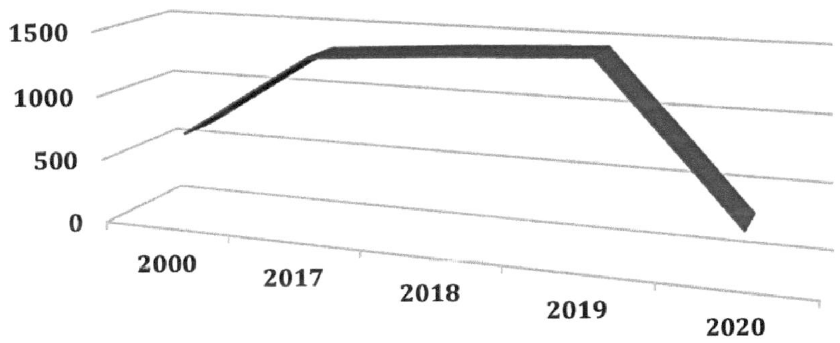

FIGURE 2.1 Tourist arrival in UAE 2020 (USD billion) [5].

The pandemic resulted in reducing international travel and made the overall tourism industry suffer along with all other industries. All the countries have faced the fatal drop, and a few are indicated in Figure 2.2.

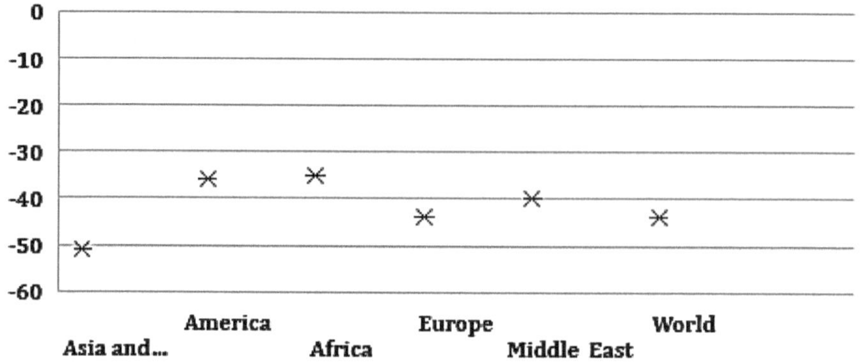

FIGURE 2.2 Decline in tourist arrival Jan–Apr 2020 vs. 2019 [6].

Several Airlines closed their operation due to domestic and international flights dropping to a minimum to zero and made most to close the business after the second half of March 2020 [6].

2.3 LIMITATIONS OF THE HOTEL MANAGEMENT INDUSTRY

The hospitality industry is currently the backbone of every country's economy with the rise of globalization day by day. People from every country visit the other to experience culture and diversity and enjoy the world's beauty. Families visit resorts and hotels during vacations to unwind and relax. Hotels are almost fully booked during peak seasons. Certain limitations need to be addressed for this industry to be considered ideal. The drawbacks are discussed in subsections.

2.3.1 LACK OF CONSISTENCY IN CLEANLINESS MANAGEMENT

The first impression of any hotel room begins with the hotel washroom. Cleanliness and hygiene are categorized as the number one preference when choosing hotels. Hotel guests who do not need many amenities would give up on WI-FI, TV, and others, but no one would choose a dirty hotel over luxuries. Having efficient housekeeping teams is very important for a hotel's reputation and reflect loyalty to the guests. One has to make sure all the rooms are clean and do not leave a chance for the customer to complain. Since the pandemic began, all world industries have been in shock, but it has dramatically affected the hospitality industry. Most hotels were assigned as COVID-cure centers, which disrupted their standard routine mechanism. The guests living in the suites had to leave as the hotels failed to maintain their hygiene standards, and the governments felt that it was not safe for healthy people to reside in a space the same as those affected. Hence, this has left a stigma, and people think twice before making their subsequent hotel reservations.

2.3.2 DIFFICULTIES IN MANAGING ENERGY RESOURCES

One of the hotel industry's biggest challenges is managing the rising expense pressures due to energy costs. Hotels are being hit on two fronts: Wholesale energy costs are expanding, principally determined by the taking off the cost of oil, while non-ware costs are additionally rising strongly. This implies it is essential to discover moderate and imaginative better approaches for lessening energy utilization. The test is to accomplish this without redirecting capital spending that is reserved for improving offices and making more appealing client contributions.

As an energy-escalated business area, the lodging business has proactively sought energy proficiency openings. Various representatives revealed that they had recuperated energy investment funds from introducing LED lighting, with some utilizing consolidated warmth and the ability to create a minimal expense, low carbon warmth and force. Lodgings will, in general, use undeniably more energy for lighting, warming, and cooling than different areas. Like this, our underlying site overviews will generally begin with evaluating the expected recompense of lighting redesigns, checking control frameworks, and investigating the proficiency of boilers and cooling gear [6].

2.3.3 NO ADEQUATE SECURITY AND SAFETY

Security and Safety is a major factor in a hotel's reputation. Guests would feel more comfortable if a home-like experience is provided, but most hotels fail to do so. They do not get facilities on time, or their privacy gets invaded. This mostly happens for famous people and known personalities.

Journalists, fans, and paparazzi constantly invade their spaces. Even in smart hotels, the system is only a step away from being breached as network credentials are all at large.

2.4 TECHNICAL ANALYSIS FOR SMART HOTEL MODEL

Before coming to a final solution, the massive data block can be analyzed with Big Data analytics in Hadoop or a similar environment. The proposed solution is based on Hadoop [7] in this chapter. Earlier, associations used to purchase costly equipment to accomplish high accessibility. Hadoop as it does not depend on equipment; however, it accomplishes high accessibility. HDFS (Hadoop distributed file system) utilizes expert slave engineering. Each bunch has a NameNode for dealing with the record framework activities and supporting DataNodes for overseeing information stockpiling on singular registering hubs (generally, there exists one DataNode per hub in the Hadoop group). HDFS stores information in records separated into at least one portion and put away specifically DataNodes.

2.4.1 PARTS OF HDFS ENGINEERING

NameNode is the ware equipment that deals with the metadata and goes about as the expert worker. NameNode is answerable for giving customer admittance to records, dealing with the document framework namespace, directing record framework activities like renaming, opening or shutting records. DataNode is where the information dwells for performing peruse or composing tasks. Given the guidelines given by the NameNode, DataNode performs block creation, block cancellation and replication [8].

2.5 FRAMEWORK OF THE SMART HOTEL MODEL USING HADOOP

The hotel system is an integrated smart device network that provides convenience (Figure 2.3). Issues of sanitation will be resolved using automated sanitation facilities. The guest's arrival time will be calculated using Hadoop and predicted to sanitize premises when the guest leaves the room.

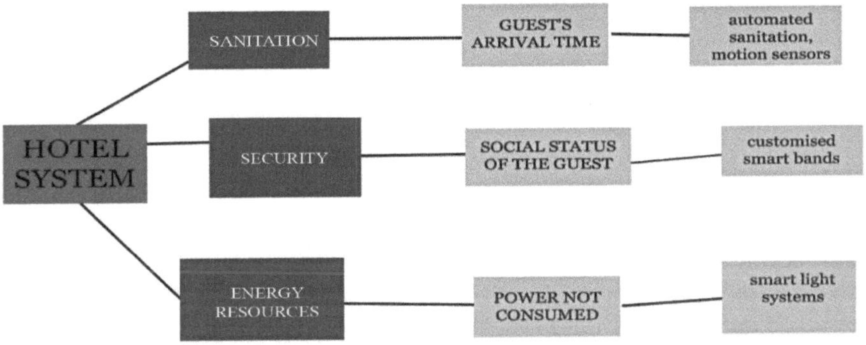

FIGURE 2.3 Smart hotel with integrated smart devices.

Security will be provided based on the social status of the guest. Customized smart bands will be given to each guest with a special configuration to access different floors as per their reservations. Lastly, power will be conserved using smart taps and bulbs. Power not consumed will be detected by the system.

2.6 ALGORITHMS AND WORKING COMPONENT

Figure 2.4 explains the concept and implementation of the proposed model, with three verticals sanitation, security, and energy conservation. These three verticals collect the data from the different entry points. The collated data is translated into components to lead to a focused and more innovative approach for smart hotels.

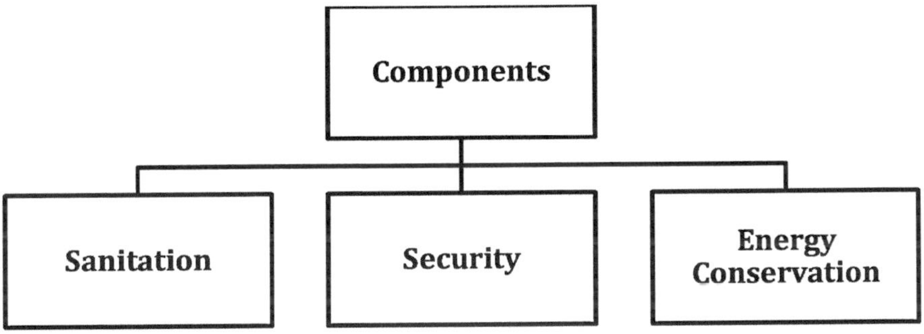

FIGURE 2.4 Components in proposed smart hotel.

2.6.1 SANITATION

For maintaining sanitation, automated sensors will be installed in every corner of the room, right where more crowd and contamination is expected, like the furniture and foyer of the room.

Waspmote sensor network can be used to establish a network and connect the IoT things, which is the automated sensor. Waspmote is a universal hardware platform for the IoT because it is open and modular. It can integrate any sensor, use any communication protocol, and send retrieved data to any Cloud system for analysis or storage. Many organizations have simplified the integration of leading Cloud software platforms to its Waspmote wireless sensor network, reducing time to market for new IoT solutions (Figure 2.5) [9].

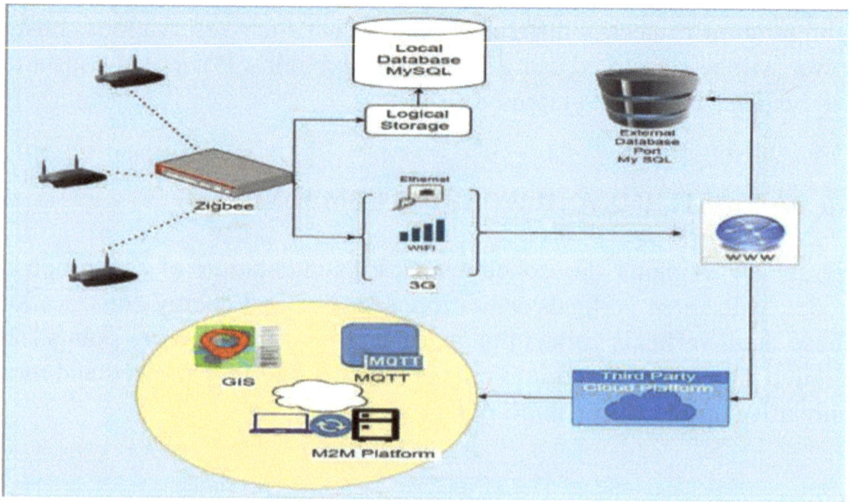

FIGURE 2.5 Sample of integration of cloud software.

Each sensor is connected to a sprinkler that is connected to the ducts that carry the sanitization fluids. Once the guest leaves the room and the door is locked, a motion sensor near the door detects motion, enabling all the sanitization ducts to open. Thus, the room gets sprinkled with sanitization fluids, which ensures no possible chance of contamination remains when the guest returns. Devices are connected using the Meshlium switch device. Meshlium is the IoT Passage to interface any sensor to any cloud stage. Meshlium is the most suggested choice for open-air networks since it is encased in a rough, waterproof walled area which shields it from the harshest conditions. It is a Linux-based switch, absolutely measured and uncommonly intended for unforgiving conditions without bargaining adaptability in the establishment.

2.6.2 SECURITY

Security in the premises can be achieved by calculating data coming from all the guests. Due to the pandemic, human workers cannot contact the incoming guests. This is why doors and smart bands will be more feasible to use. Customized Scrum management will be more feasible to adopt in this kind of cases.

Various methods have recorded all requirements. An obtrusive method for data collection, such as an interview, has been adopted here. Interviewing people in more technical jobs/supporting IT infrastructure has been given the preference. After analyzing the visible data requirement, lock unlocking the door with QR (quick response) code, maintaining the log of each entry and exit, authorizing, and authenticating key, and providing the entire interface of key management (Figure 2.5).

Few more requirements are also important to notice and include in the system, such as after a successful scan of the QR code, the system must lock/unlock in a few seconds (as defined by the administrator). If the QR code is invalid, notification of the same needs to be communicated immediately (Figure 2.6).

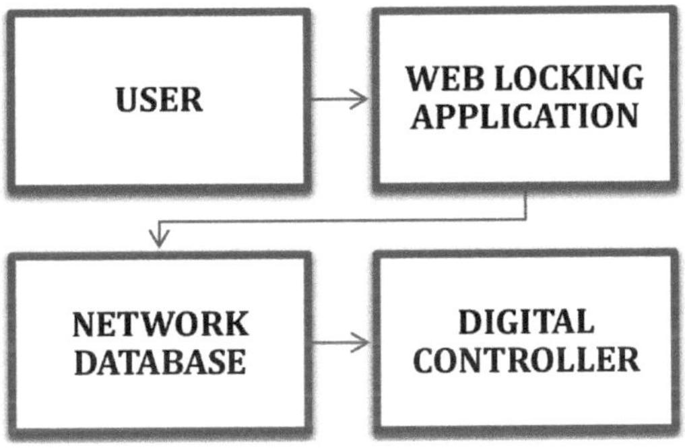

FIGURE 2.6 Client server-based architecture for digital lock.

Authors [10] explained an interesting concept of a mechanism, which can be developed with the help of an Arduino Uno microcontroller, along with the support of a camera and QR code. This can be managed with Android Studio IDE. Data generated by this system (Lockmate) can be stored in server for the security and safety of data. Each guest is given a smart band with a QR code, SOS button and an LED light with a sensor. The smart band is the ticket for the customer and will be used to link them to the central database. All their vital information will be entered into the MYSQL database once they enter the premises and scan their band.

2.6.3 ENERGY CONSERVATION

For energy saving purposes, motion sensors with smart lights will be used that are made with Arduino and PIR (passive infrared sensor) sensors. Those sensors can assist with evaluating the movement happening outside the study halls. We utilize a PIR to find activities outside the study hall. That gives us a 0 or 1 yield. The event that 0 happens implies that no movement is identified, or on the off chance that it gives 1 implies that movement is recognized outside the study hall. So, if the climate is shady in the stormy season and dimness is there, the lights will be ON naturally. The entire design on the Arduino will be finished. Arduino is a gathering for open-source hardware. Both the equipment and the product utilize it.

Arduino sheets utilize a sensor, a finger on a switch, to peruse input lights [11, 12]. An inhabitant's sensor is a control framework that detects inhabitants and the number of individuals inside the room, and Arduino gets the sign from it. The specific individual included will be reflected in the LCD board. Sensors are not serviceable whenever, so enormous sensors help must be forestalled for any mistakes. Circuit work is fundamental. The fan starts running sluggishly when the joined roof ventilator [13]. Whenever a guest leaves the premises, the motion sensors detect no activity in the room, and hence the lights switch off. When the guest returns, the lights switch on as constant motion is detected.

2.7 WORKING OF THE COMPLETE SYSTEM

When customers 'A' want to make a reservation, they will make bookings ahead of their check-in date through any online portal. The user receives a bill with a QR code to show to the camera. Based on their reservation choices, they may or may not be given access to the entire floor where high-class security is provided. 'A' will be given a smart band with the same QR code that they are required to scan at the main entrance to gain access using the Image Recognition Algorithm. Their name and details are entered in the database, permitting residents of the hotel premises. The data is then transferred to the hotel's private cloud network via the router and switch, activating their smart band. They are given their room/rooms as per reservation and facilities like every other hotel. When 'A' leaves the premises, the motion sensors detect movement using Infrared Detection Technology to detect the presence of humans. When the output is returned as false, the lights switch off, and the sprinkler ducts open, sprinkling sanitization fluids in the premises. When the guest enters their room, their room is already sanitized and cleaned.

The situation is similar for the exclusive floor, but there is extra scanning for precaution (Figure 2.7). Only people with smart bands having corresponding QR codes can enter the premises to avoid any form of trespassing. Rest all remains the same for both floors, except for a few minor changes.

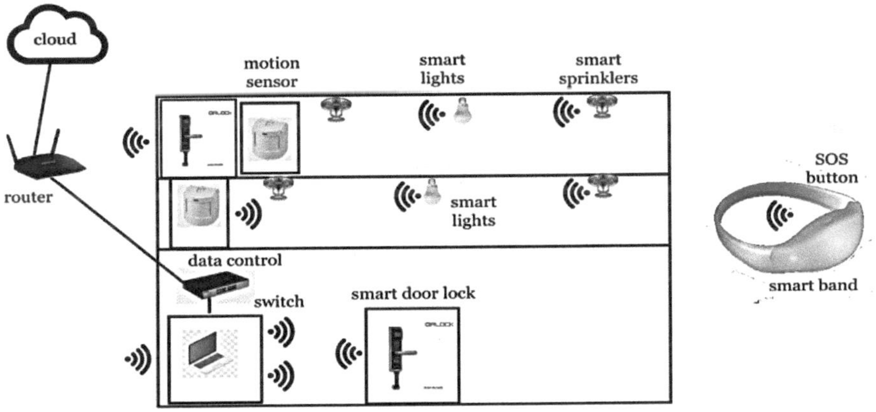

FIGURE 2.7 Concepts of contactless data capturing.

2.8 RESULT AND ANALYSIS

A public survey was conducted, answered by close to 500 respondents. They were presented with the model and the hotel's features, out of which 95.5% of people agreed to add smart bands for better access to the facility and ease of operations (Figure 2.8).

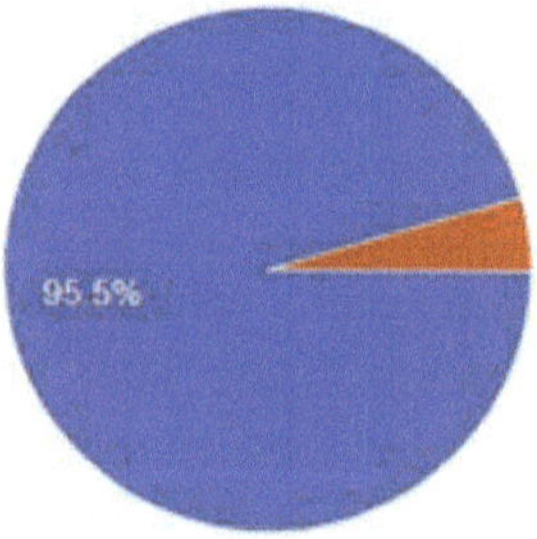

FIGURE 2.8 Smart band for all guest that have SOS signaling.

Around 86.4% (Figure 2.9) agree to the entire floor plan for known personalities and celebrities, as trespassing and invasion of privacy-based incidents are becoming more and more common. Around 81.8% (Figure 2.10) of people also preferred the automated sanitation facility plan over standard sanitization plans as less human contact was involved and would be more consistent. They said it would be helpful during the COVID-19 pandemic as regulating the sanitization of every individual is very difficult and requires much labor.

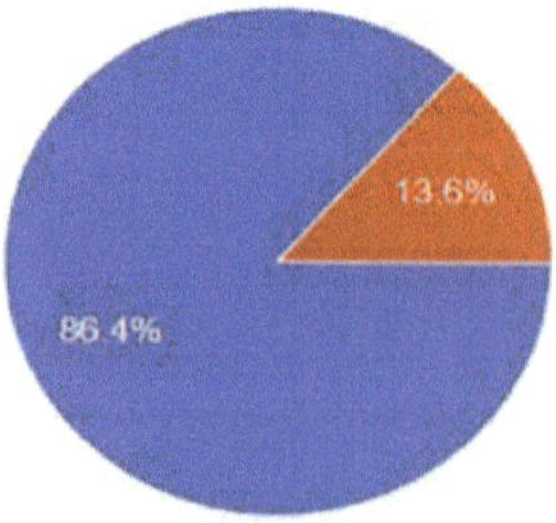

FIGURE 2.9 Special floor for famous people.

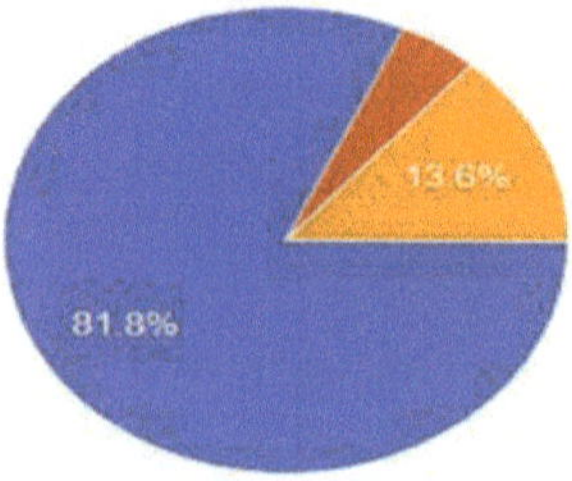

FIGURE 2.10 Automated sanitization facility.

2.9 CONCLUSION

The hotel industry is a growing field and will flourish well when led in the right direction. In the 21st century, everything is becoming automated due to the rising popularity of the Internet and new technologies. Hotel management can effectively combine IoT to establish a safe industry, especially during the pandemic, when everything is going digital. The proposed system will be preferred by common people, as it provides smart solutions to common but not much talked about issues that customers face. Using smart gates with QR codes will help evict any invaders or passersby. Sanitization of the entire premises would make a guest feel safer when they return to the room, and the exclusive floor would make any celebrity comfortable, which is not the case most of the time. They face much exposure to the media even if they do not want to. Stalkers are around them all the time. We can keep anyone suspicious away from them using smart band technology and protect their image. The smart bands also help regular guests feel safer as they are equipped with an SOS button in emergencies. Implementing this system can make anyone feel at home, even if they are miles away from it.

KEYWORDS

- **big data**
- **cloud computing**
- **divide and conquer**
- **energy resources**
- **hotel management**
- **image recognition**
- **sanitation**
- **smart hotel model**

REFERENCES

1. Martin, W. *Five Energy Challenges the Hotel Industry is Experiencing.* https://www.centricabusinesssolutions.com/blogpost/five-energy-challenges-hotel-industry-are-experiencing (accessed on 07 September 2023).
2. *Impact of the Pandemic on Tourism* – IMF F&D. https://www.imf.org/external/pubs/ft/fandd/2020/12/impact-of-the-pandemic-on-tourism-behsudi.htm (accessed on 07 September 2023).
3. Melissa, R. *5 Recession Lessons Hotels Should Remember in 2022* | Hospitality Net. [Online] https://www.guestcentric.com/?utm_source=HospitalityNet&utm_medium=October2021&utm_campaign=HospitalityNet&utm_id=5RecessionLessonsForHotelsArticle (accessed on 07 September 2023).
4. News Article | World Travel & Tourism Council (WTTC). https://wttc.org/News-Article/WTTC-identifies-digital-solutions-for-governments-worldwide-to-significantly-restore-international-mobility (accessed on 07 September 2023).
5. International tourist arrivals decreased in January–April 2020 compared to 2019 by regions (%). | *Humanities and Social Sciences Communications*. https://www.nature.com/articles/s41599-020-00630-8/figures/1 (accessed on 07 September 2023).
6. Mulder, N., *The Impact of the COVID-19 Pandemic on the Tourism Sector in Latin America and the Caribbean, and Options for a Sustainable and Resilient Recovery* (p. 49).
7. Evans, C., (2013). *Big Data Storage: Hadoop Storage Basics*. Computerweekly.com. Computer Weekly.
8. Iconiq. *Hadoop Explained: How Does Hadoop Work and How to Use It?* [Online] https://www.dezyre.com/article/hadoop-explained-how-does-hadoop-work-and-how-to-use-it-/237 (accessed on 07 September 2023).
9. Libelium. *Meshlium*, [Online], https://www.businesswire.com/news/home/20131118005374/en/Libelium-Sensors-Connect-with-Cloud-Platforms-for-the-Internet-of-Things (accessed on 07 September 2023).
10. Ahmad, R. A., & Masitah, G., (2017). Lockmate: Digital door lock system using QR code. *UTM Computing Proceedings Innovation in Computing Technology and Applications* (Vol. 2)| |ISBN: 978-967-0194-95-0.
11. Anil, A., Shukla, V. K., & Naranje, V., (2021). Tracking vehicles through GPS module and Arduino UNO. In: *2021 9th International Conference on Reliability, Infocom Technologies and Optimization (Trends and Future Directions) (ICRITO)* (pp. 1–6). doi: 10.1109/ICRITO51393.2021.9596167.
12. Altaf, H. F. H., Shukla, V. K., & Tripathi, A., (2021). Sorting of objects from conveyer belt through color detection and audrino UNO. In: *2021 International Conference on Communication Information and Computing Technology (ICCICT)* (pp. 1–5). doi: 10.1109/ICCICT50803.2021.9510037.
13. Anurag, Y., & Gayathri, N., (2020). Application of IoT with motion sensor for smart learning environment. *International Journal of Recent Technology and Engineering (IJRTE)*, *9*(1). ISSN: 2277–3878.

CHAPTER 3

Blockchain Integration in the Hospitality Industry

WONDA GROBBELAAR,[1] AMIT VERMA,[2] and VINOD KUMAR SHUKLA[3]

[1]*NSRIC, Canada*

[2]*HTMi Switzerland, Dubai, UAE*

[3]*Amity University, Dubai, UAE*

ABSTRACT

The hospitality and tourism industry loses billions to non-financial fraud yearly and is the highest target sector for cyber security. It increases the overall cost, but it also harms companies' reputations and creates a negative working environment for employees. Artificial Intelligence (AI) and Blockchain is frequently used successfully in different business sectors and can improve the workflow in the hospitality sector. The technological possibilities of AI and Blockchain can offer a secure option for document fraud, currency exchange and identity theft, only to mention a few. The fundamental principles of blockchain technology and AI will be covered in more detail in this chapter, with an emphasis on the elements that are crucial to understanding the rest of the chapter. It will provide knowledge that academics and professionals will need to better comprehend blockchain technology and apply it to the travel and tourism industry. It will, however, also draw attention to flaws and crucial study areas. An examination of the Blockchain's potential to disrupt the economy is covered in several ways that could change current market structures and possibly result in the disintermediation of existing players in the tourist sector and the creation of new ones.

3.1 INTRODUCTION

The hospitality sector loses billions each year through several prospects to feat omitted or blemished anti-fraud controls in this industry [1]. The sector includes tourism destinations, attractions, different accommodations, transportation, F&Bs and multiple other ancillary services. Non-Financial Fraud, including employment credentials internal and external documents, contribute to high losses. It costs the industry much money, but it hurts the companies' reputation and creates negative employee morale. Artificial intelligence (AI) technology has created transformational change across different sectors, and the endless potential uses have created already practical applications in the hotel industry [2]. Timesheets filled in with the wrong information, fraudulent qualifications, and false reports can contribute to huge losses.

The information is shared between different companies within the travel industry for the smooth operation of a traveler. Considering the amalgamation of several ancillary services and retail service providers for every step of a hospitality and tourism operation, the industry is supposed to be significantly impacted through the positive implementation of AI and Blockchain technology. Blockchain is suitable for enhancing the outcome within business scenarios where numerous service providers collaborate or participate in a transaction. The usage of Blockchain technology is implemented to remove the middleman involved in hotel bookings or gain loyalty program points from the transaction through the usage of distributed ledger technology. The technology can upsurge the innovative business models through its three distinctive, impressive features of Transparency, Security, and Immutability to not only improve the workflow but also enhance safety. For instance, an agent communicates the passenger information between several other service providers and further instigates the possibilities of sharing and tracking the information.

The technology can establish the responsibility of accessing and storing crucial information across the network with ease and reliability. In terms of the disadvantages, one of the parties in the blockchain needs to have access to the internet connection. Not all countries around the world have stable network connections. Low bandwidth, the quality of service and the type of network architecture can influence the success of a company's results when using blockchain. Upcoming broadband, for example, 5 G, might contribute to the success of blockchain but is limited to the type of device and the country it is being used in. Building an effective network is expensive, and it might still be a while before all countries have a stable network and 5 G broadband.

The non-centralized approach of blockchain technology plays a vital role in controlling the access of information by reducing the chances of loss of data by any unintentional deletion or intentional hack or cyber-attack by ensuring the transactions are always traceable due to the apparent reason of never being offline in the chain [3].

Blockchain spending on different applications for 2020 is illustrated in Figure 3.1. "According to a recently published report by IDC (International Data Corporation), Global expenditure on blockchain solutions is projected to be $4.1 billion in 2021, indicating the upsurge of 50% equated to 2019" [4].

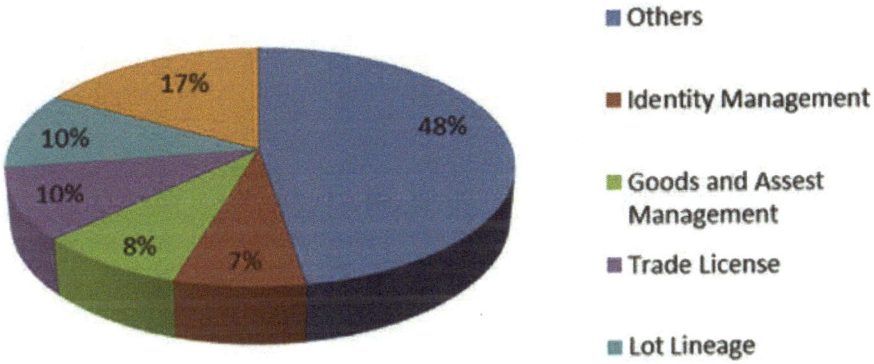

FIGURE 3.1 Blockchain spending guide for 2020 [4].

As stated in Figure 3.2, the banking sector will continue to introduce more blockchain applications to the sector. Deutsche Bank, Standard Chartered and Westpac are well-known names that are already using blockchain. Cross-border payments are part of several applications, lower processing fees, trade finance and an easier way to process loan applications. One of the biggest growths in blockchain spending will come from manufacturing and professional services. Process manufacturing creates items by combining raw materials and ingredients using a recipe or a formula.

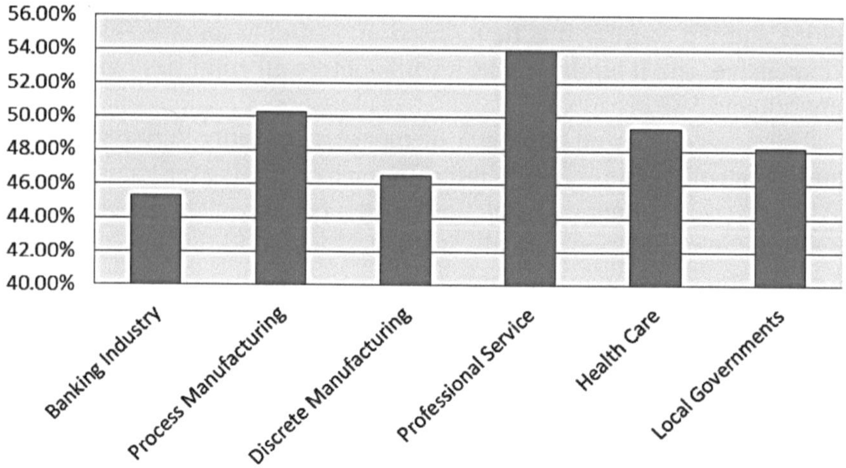

FIGURE 3.2 Blockchain trends in industry (2021–2025) [4].

The hospitality industry, especially the F&B sector, fall under this sector [2]. Discrete manufacturing is the manufacturing of distinct items, for example, aircraft. The highest spending growth will come from professional services. Professional services include lawyers, teachers, and accountants and a degree or accredited qualification—non-financial fraud, for example. The rightful utilization of blockchain in the hospitality industry's human resource department and educational institution may combat forged qualifications/certification, etc., through blockchain technology and substantially transform the tourism industry.

3.2 BLOCKCHAIN IN THE TOURISM INDUSTRY

New innovative business models-built customer relationships and facilitated new forms of inter-organizational collaboration. The new business model in the hotel industry is developing end-to-end services from looking information-seeking, including services like gift deliveries babysitting, till checkout. One such model is multisided platforms. The platform combines product services and technologies that develop the forum to initiate the direct interactions between customers and participating clusters to enhance the value [5].

Blockchain will revolutionize business models in future. The travel industry is one of the highest target sectors for cybersecurity breaches and fraud attempts. Each of the offerings on online platforms comes with a unique risk, paving the way for an increase in all types of fraud. The overall appearance of fake travel agents is much more sophisticated today than the average fraudster. Travel agents use 1–2% of their yearly income to combat fraud. The average price of a fraudulent travel booking can cost anything from $283-$588. Often the traveler is unaware of the propensity of frauds in the travel industry, for example, transaction issues [6].

Loyalty fraud, identity fraud or content abuse such as fake reviews on accommodation are on the increase, and often the traveler is unaware of this criminal activity. Blockchain offers a secure option for currency exchange or credit card transactions and facilitates automated and programmable micro-transactions, which are decentralized.

With the increased online usage and throbbing technological influences, Consumers expect real-time payments. They do not want to wait any longer than a few days for payments to go through the different channels, which is smoothly possible through the blockchain for enhancing the service standards of overseas transactions through the safety and timeliness of the innovation. Blockchain uses proof-of-work consensus algorithms. Implementing and managing a blockchain project is very difficult. All staff members need to be trained on blockchain and utilize it successfully. It requires in-depth knowledge from businesses to implement bitcoin successfully. Staff must be educated on utilizing blockchain and then ensure that the management team can understand the complexities of a blockchain applied in a business. They must recognize the warning signs and how to avoid falling prey to attacks.

3.3 BLOCKCHAIN AND ITS OPERATIONAL UNDERSTANDING

Blockchain technology is considered crucial to impart structural changes and innovative business models due to its imperative characteristics of safety, transparency, and network-based security. However, the impactful utilization of the technology needs to be translated in the context of the Hospitality and Tourism industry. This would project the novel framework of the industry through the emergence of technology-based services and aligned products to derive the core values hidden in the consumer's demand.

The operational understanding of the blockchain would be a foundation through implicit examples and suggestions to boost the possible visibility and insight of the industry.

The technology is based on the chain's blocks (records of each transaction) (nodes in a network, including each peer-to-peer node network). Each transaction in, between, and the node is recorded and accessible through the "chain" public network database. The digital ledger is the "storage room." As depicted in Figure 3.3, the Practical blockchain applications include tracing food items from farm to fork, record keeping of academic qualifications, and record-keeping of customer information, such as allergens, preferences, and requirements. New entries by participants can be seen in real-time. Through a special network, every computer in has access to shared records. Data is updated live, making it almost impossible for fraudsters to tamper with transactions. Computers verify transactions in blocks. Contrarily to previous belief, blockchain hacks have increased since vulnerabilities were discovered.

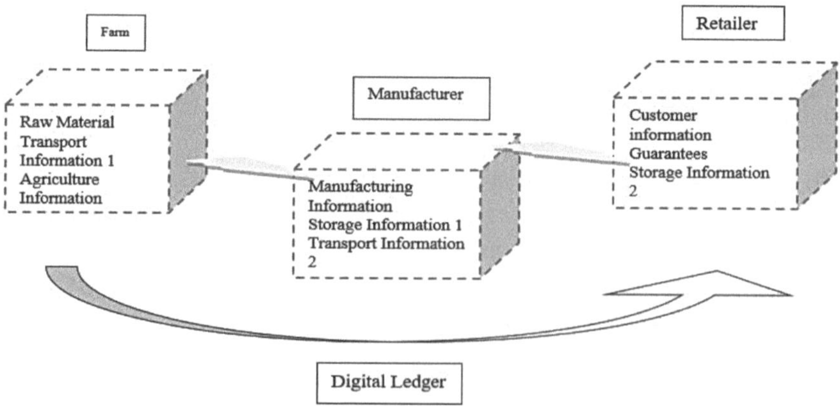

FIGURE 3.3 Basic illustration of blockchain.

For blockchain to be successful, good teamwork is very important. As part of the service sector industry, anti-fraud contingency plans are essential to the industry's success. Share relevant information with best practices and putting knowledge into action is a key factor to the future success of this industry. Blockchain is an ideal way to prevent fraud in many areas. The ledgers can be shared in real-time and only updated once all parties agree. Blockchain can be divided into a public or private blockchain. A public blockchain has an open network, and everyone has information. A private blockchain is an invitation-only blockchain governed by a single entity.

Blockchain has many advantages that apply to many industries, including tourism and technology. To understand blockchain, you need to understand a simple form of data structure. The data structure of blockchain is a back-linked list of blocks. Nodes make up link lists.

The blockchain algorithms are designed to assure the data's safety by the adjustments to the blocks enabled by the global data, preventing failure, which means there is no single failure point, and the data gets through the nodes to make up the blockchain's ledger. Moreover, the chain of trust guarantees the way that data is stored on the blockchain has not changed due to its decentralized approach of nodes that do not allow any single unit to take complete control at any point of transaction, leading to the development of blockchain with distributed control on the processing of information.

Blockchain technology combines mathematics, cryptography, algorithms, and distributed consensus algorithms. Blockchain can assist with tracking baggage or finding the baggage location after being handed over to the airline staff at the airport counter. The checked-in baggage within any airline passes through several airport staff and department controls in a single passenger journey enhances the probabilities of errors and omissions that result in lost or misplaced baggage [7]. Blockchain will be able to track the baggage of the customer. The global access to the data in real-time would translate into significant time and profit mileage to the organizations by reducing the gap between the consumer and service provider by ceasing the individual entity's influence in any operation or transaction.

3.4 NON-FINANCIAL FRAUD IDENTIFICATION

Fraud can be disguised in many ways and cost the hospitality and tourism industry millions of dollars in a year. One individual person should not manage control of any department. Non-financial fraud can include asset theft, falsifying documents such as academic qualifications, progress reports, false reviews of a service provider, and corruption in kickbacks to staff members [1]. The nature of certain crimes makes it hard to identify them immediately, and it might take time before it is exposed. Fraud incidents in a company will damage your reputation. Fraud can be in the form of booking scams, or some of the booking scams can include hidden fees that are added ambiguously to the customer's bill, non-descriptive language to advertise a room in a hotel or pictures from sneaky angles to make the room look bigger. The organization must establish equilibrium in achieving the service standards of the online guest experience and ensuring the safety of the transaction and data protection of both property and guest to not hamper the reputation.

3.4.1 EMPLOYER CREDENTIAL FRAUD

Fraudulent qualifications and false reports can harm a company in different ways. Fake universities sell certificates for all degrees, and staff members get employed based on these qualifications:

- On an annual basis, more than 200,000 fake degrees are awarded to students and keep increasing every year;
- A fake degree will cost anything between $500 and $2500;
- Currently, there are more than 370 educational websites that offer fake degrees;
- Many job applicants lie on their Curriculum Vitae (CV) about their education.

Not only are fake credentials increasing, but candidates who already earned a credential may also falsify their academic scripts. Hiring those employees with fake qualifications on their CV's can be a costly affair for any company and often result either in additional training costs to get the employee to the level that is required to do the job successfully or a high turnover of staff [8].

Another common issue in the service industry is fraudulent work history. Most of the time, candidates with a fraudulent work history cannot meet the required standards of the job descriptions and sometimes, the company needs to hire additional staff to complete the task.

Inconsistency on staff creates an unstable working environment and costly legal action against these fraudsters. In the past, it was believed that fake universities sold these qualifications; however, these days, a degree can be bought from a legitimate university [8]. Many students get caught in networks that offer help with assignments at a fee. The student from the same university will recruit possible candidates amongst his friends for these fake degrees and make some extra money out of this.

The lack of staff and a heavy workload prevent an HR department from conducting a proper credential check on candidates. These types of non-financial fraud have increased during a pandemic, and robust credential checks have become more critical than ever. The negative side is that these reference checks are time-consuming, and it is not always reliable. Placing these degrees in a block will reduce work history fraud and eliminate employing staff with a lack of knowledge and work experience has impact on the net profit statements of any company. According to Brown, the rising demand for degrees around the globe has increased the fraud related to degrees to satisfy the requirements of qualification [9]. This has given birth to a billion-dollar industry and has impacted all nations from its ill impact [10].

As per the recent US data, 6% of bachelors and 35% of associate degrees are forged [8]. Most employers encounter the problem. It is found that either these degrees are produced through an unrecognized fake institution or due to the fraudulent activities of the student or due to the corrupt officials in an appropriate institution [11].

In such a situation, the time, energy, and effort of a student, institution, or the organization's human resource department is wasted. And the blame must be claimed to the non-transparent system of authorization of institutions, the authenticity of the degrees and the validity of the issuance of certification to the human resource department students and their parents if deeply integrated. This technology may provide access to factors that may calibrate the operational excellence of an organization by seeking reliable information of the actual hours of practical training, resultants, and attitude of the student, etc., which may be supported with the degrees and certifications.

This will ease the authenticity of the degree and reduce the struggle of finding the appropriate stars of the organizations.

3.4.2 DOCUMENT FRAUD

Any document that is changed or modified to create a different reality to the receiver is described as falsifying documents. Some examples include timesheets filled in with incorrect information but not limited to additional working hours added or signing in by a friend for someone not at work.

Examination papers that leak before an exam, fake medical certificates to justify the absence from work, and false reports on work-related issues can contribute to financial losses.

3.4.3 IDENTIFICATION SECURITY

Blockchain can be used to identify tourists and guests in the hospitality industry. During the immigration process, a tourist must be identified. Blockchain is regarded as a secure technology; if it is executed appropriately, it may eradicate current identity issues. Often there is no standardized way to identify a tourist in two countries, and it opens possibilities for identity fraud in different ways. Digital identity and Blockchain can better control a tourist's own identity. From the moment the passenger embarks in his home country on the journey till the destination, personal identification is several times required.

The passenger's identification in the hospitality is a significant aspect easily manageable through the Blockchain. Blockchain is regarded as a secure technology. Digital identity can be the requirement of the validation of each passenger through the identity is a mandate to contribute to better control over a tourist's identity. The adoption of the Blockchain could allow for a global digital database, with passenger's biometrics recorded and stored in a secured distributed ledger. Examples of biometrics include facial, voice recognition and fingerprints. No waiting time is involved in verifying a potential customer. Once the information is stored in a distributed blockchain ledger, the data will validate customers' identities without repetition. This app can be first runner in the personalization. For example, if a hotel wants to reward a customer for his loyalty, the offering can be sent to a specific room.

Personal identification can also be used when the guest checks in at the hotel. The guest walks up to the check-in desk and enter his reservation number An OTP (one-time password) is sent to the preregistered mobile number of the guest in the form of a WhatsApp or SMS. The guest will then enter this code to verify his identity. The check-in desk is connected to a blockchain platform that will assign a room to the guest. A time-sensitive token will be generated. The token would then assign a room according to the guest's preference. It will allocate a room according to the guests' preference; also, it can be extended to send to inform the kitchen about allergens that might occur, and meals can then be prepared according to the guest's taste and health. The smart lock will expire on the last day of the reservation at a set time. Life applications are already happening in the sector. The company is investigating the application of single travel tokens, which Blockchain will authenticate to minimize document checks during a journey.

As illustrated in Figure 3.4, identity theft is the highest within document fraud. Practical solutions to combat identity theft exist and are very effective; however, the disadvantages include a very high cost per document and the time it takes to verify the identification document. In the future, implementing the blockchain in the multifaced organizational structure could allow for a shared digital database, with passenger's biometrics recorded and stored in a secured distributed ledger that can be assessable for all relevant parties involved. Examples of biometrics include palm veins, facial and voice recognition, and fingerprints. There are many unauthorized immigrants around the world without any identification.

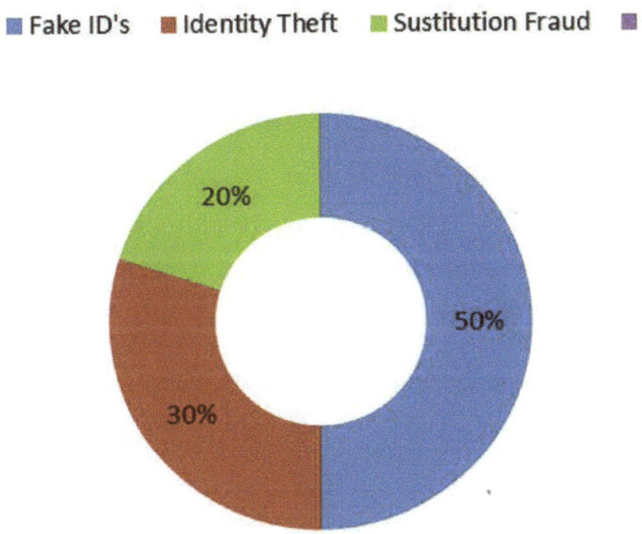

FIGURE 3.4 Different types of identification fraud [1].

The refugees total by the end of 2019 were 26 million, as depicted in Figure 3.5. Half of the refugees are very young, and few due to the varied status or not determined status for an asylum seeker under the international protection but not identified as a refugee. Some has lost their identification document through difficult circumstances, while others destroy their identification on arrival in a new country to avoid deportation. By using biometrics and blockchain, faster solutions can be found for identification. No waiting time is involved in verifying a potential passenger. Once the information is stored in a distributed blockchain ledger, the data can validate passengers' identities without any repetition. This application can also be applied for personalized marketing. For example, if a hotel wants to reward a guest for loyalty, the offering can be sent to a specific room.

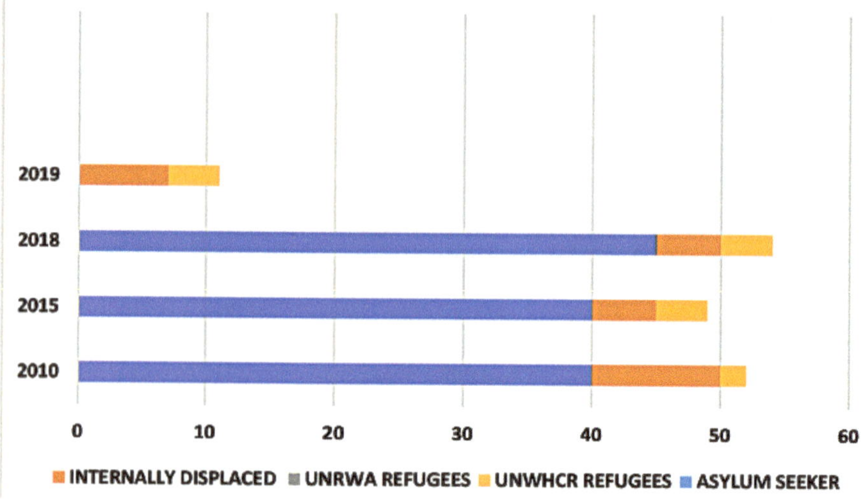

FIGURE 3.5 Global forced displacement worldwide by the end of 2019 in millions [12–14].

Blockchain can also be applied to prevent identity theft. For example, the fraudster may use a person's credit card to pay for a hotel booking. In some countries, family members will use fake identification to receive financial benefits from the government even after the person's death. Terrorists can use fake travel documents to carry out their illegal activities.

3.5 EFFICIENCY OF BLOCKCHAIN

In the Blockchain, a middleman's involvement is required in any process. Blockchain creates faster transactions by a cross-border method. By removing third parties, any company can significantly reduce the overall cost as no additional payment needs to happen. Blockchain can also assist with creating a lean culture in companies by streamlining the supply chain and removing all the wastages as described in Lean Six Sigma [15–17]. Blockchain will enhance the food industry's efficiency, transparency, and traceability. The ability of tracing the diet items in a HACCP (hazard analysis critical control points) certified kitchen is very important and forms a part of the requirements to implement the certification successfully. In the past, records were complicated to keep as the information around farming and where the food came from was, in most cases, mediocre. It was also time-consuming to keep a proper traceability log of every food item the purchasing department bought. Guest requirements have also changed and are more informed today. Guests require transparency around the food they consume today, for example, the distribution chain of food and ingredients. Blockchain can store information from farm to fork, and the digital ledger can be shared with all the required parties involved.

3.5.1 LOYALTY PROGRAMS

Loyalty programs are very popular in the travel and tourism sector to attract customers and let them come tomorrow. Blockchain can be a platform where this data can be stored securely and allow customers to check the status of their points at any given time. Limitations could be added, for example, expiry dates when loyalty points. The customer can exchange these loyalty points for the cryptocurrency. Blockchain can assist with the minimization of the ongoing fraud in the loyalty programs and enhance the value and quality. Loyalty programs can offer rewards to its digital identity holders and those. Combining loyalty programs across different businesses challenges will be beneficial for business if the potential customer can use his loyalty points to buy products at the airport mall or apply them at the hotel where he or she is staying. Lost business opportunities will be minimized, and the customers' overall experience will be enhanced.

Blockchain technology with integration of AI can offer customers a solution to use loyalty points earned from air travel to shop at airport malls, rent a car, or hotel. This will assist the customer experience and enhance the business opportunities by combining multiple partners in their membership/loyalty program depicted in Figure 3.6. Multiple travel companies may redeem the points against the accrued points mileage based on the Blockchain distributed ledger to lead the multiplication of the redemptions in the industry. It will also create the option to transfer loyalty points to friends. By incorporating Blockchain, Dubai holdings have signed a deal for a project to incorporate "the internet of loyalty points, promotions, and rewards." The Jumeirah Group uses Loyyal, the same loyal platform Scandinavian local governments use to boost tourism [17, 18].

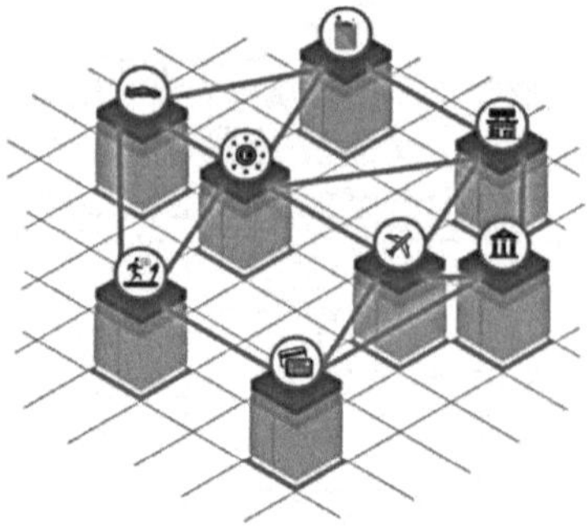

FIGURE 3.6 The loyalty program Loyyal is displayed [18].

Loyyal is an innovation group that is currently building a global loyalty operator network around the globe. Dubai aims to be the most connected and "happiest" city in the world. Smart Dubai believes that blockchain technology lacks meeting its vision and goals of the future.

3.5.2 MAINTENANCE, REPAIR, AND OVERHAUL

Through blockchain technology, the digital ledger can be shared by different airlines. Blockchain can record every part that is installed or removed from an aircraft and record how long ago a specific part was replaced. Unplanned maintenance can be avoided, and time spent on routine checks can be reduced. Information required for aircraft maintenance checks, for example, the credentials of every person who interacts with the aircraft, can be recorded and stored. Blockchain can assist in calculating the average stopovers and take-off time, and in this way, all stakeholders can get accurate runway information. Blockchain can also create value on the secondhand market for well-maintained aircraft and generally improve routine efficiency.

3.6 DISADVANTAGES OF BLOCKCHAIN

To implement and manage blockchain requires advanced skills. Staff members need to be trained on blockchain to maintain it successfully.

Blockchain is also very expensive to implement in a company. Blockchains uses excessive energy, and if the companies' goal is to create a cleaner environment, it will add to energy waste. When looking at lean principles, this will not contribute to the goal of a leaner company. Scalability remains one of the blockchains weaknesses. Once data is entered on the blockchain, it is challenging to modify. One of the parties in the blockchain transactions needs to be connected to the internet to verify the transaction. In many countries, bandwidth is unreliable, and the quality of service is inferior.

3.7 THE IMPACT OF BLOCKCHAIN ON THE TOURISM INDUSTRY

3.7.1 EARNING AND SPENDING VALUABLE MILES

Certain airlines tie-up with banks to offer 'Airmiles' points to their customers. Customers can earn miles when shopping and paying with a credit or debit card and earn Airmiles that can be collected and used at a certain point in time. Whereas if the member is not registered as a frequent flyer for the airlines, the opportunity is missed. Airmiles will expire if they are not used in 15 months. Blockchain is changing travel loyalty programs for consumers and companies. These days, it is possible to pay for your Uber ride with air miles.

3.7.2 FRAUD PREVENTION

Loyalty programs remain one of the fraud prospects due to the high monetary value of the accrued points or mileage. Sometimes, the owner of the loyalty points is unaware of what is happening to them or does not know the number of points he has available. Loyalty program fraud impacts companies in two ways:

- Airlines and hotels are losing loyal customers through loyalty program fraud; and
- Airlines and hotels must pay customers for lost points through fraud.

Using blockchain technology can minimize these types of fraud.

3.7.3 BAGGAGE TRACKING

Baggage tracking, one of the mandatory segments of the Tourism and Hospitality industry, will be energized with the implementation of blockchain technology. It will reduce the possible errors and omissions in the logistic operation and tracking. The centralized approach of blockchain will provide the accessibility of data to different agencies of operation and security. This may also foresee the possibility of automatizing the baggage support system.

3.7.4 BLOCKCHAIN IN DEPARTMENTS

Blockchain has already impacted businesses in many forms. To run data analysis in real-time and monitoring changes is a massive benefit for any company. Blockchain can help to combat non-financial crime and improve security in all departments.

In Figure 3.7, the importance of Blockchain for a hotel or a hotel group is explained. All the different departments have access to the digital ledger, containing data that one or more departments could require. Often data, for example, in the supply chain or HR department, are siloed, and data flows only linearly from one department to another. As demonstrated, Blockchain can help to break down the silo mentality and improve efficiency between different departments.

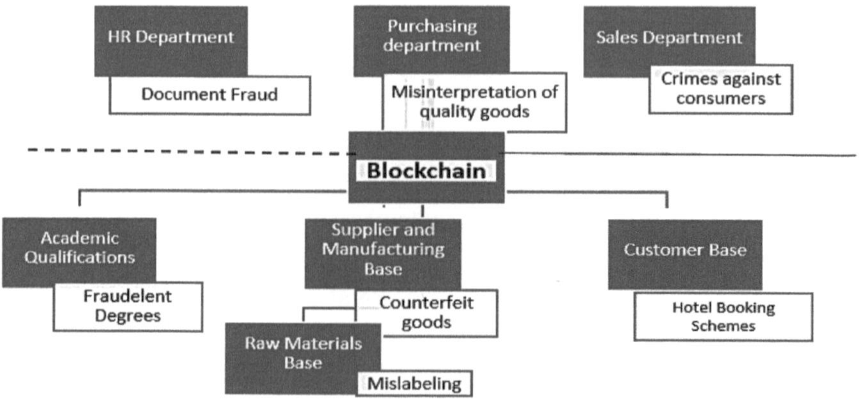

FIGURE 3.7 Potential fraud areas are identified in the hospitality sector and how blockchain assist with all departments as it is transparent in nature.

Chefs have preferred vendors that pay kickbacks when purchasing all the ingredients from these companies. Fewer products can be shipped in some deliveries than the original order; however, the total amount is charged. Often staff in hospitality are underpaid and open to all kinds of fraud. The staff will accept products that are less than the initially ordered amount. An arrangement will be made between the supplier and the staff to receive this less quantity. The supplier will pay rewards to the staff regularly. Another example will be a two-night stay booked by a customer. The fraudster will then void one night when the payment was made, write it off against an unhappy customer, and then pocket the money himself.

3.7.5 BLOCKCHAIN IN THE HOSPITALITY INDUSTRY DOMAIN

The hospitality industry includes various sectors divided into many sub-sectors, including accommodation, F&B sector, tourism, transportation, and events. Often, non-financial fraud is easily overlooked due to the complexity of the business. As illustrated in Figure 3.8, the involvement of the Blockchain is displayed in the domain.

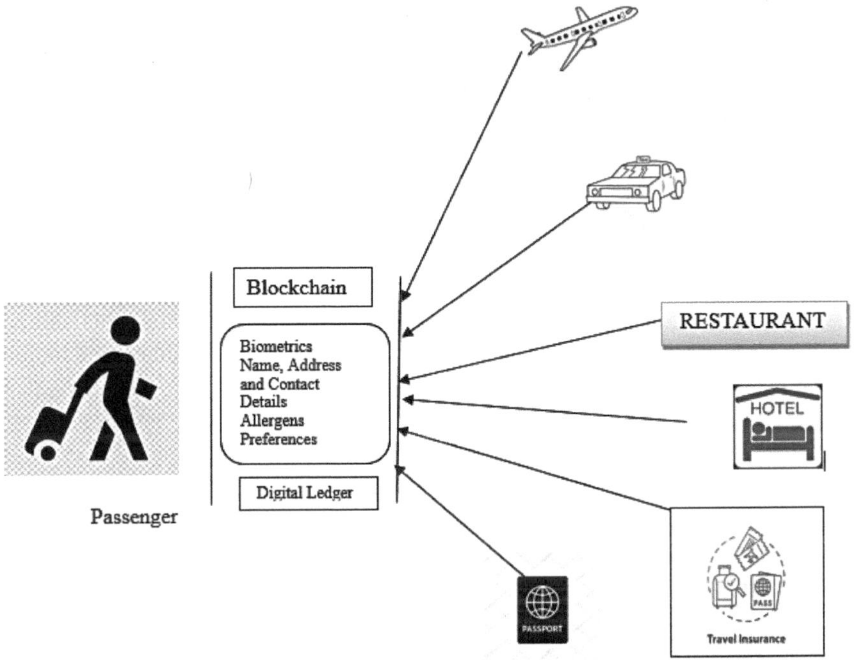

FIGURE 3.8 The different sectors and the value of blockchain is demonstrated.

All the passenger credentials are collected on Blockchain, which is accessible for the different stakeholders in a digital ledger. It might be possible in future that no longer paper copies of visas will be required, as all the important data in the form of biometrics and personal information will be recorded in this ledger around the world. Once the passenger starts the journey, he can move freely through airport security as the authentication kiosks will be verified by cross-checking facial scans with passports and palm prints. Once he arrives at his destination, a licensed taxi with an accredited driver can be used. The taxi will know precisely when the passenger arrives and his preference for a car. The restaurant will know before the time of arrival what the allergen of a guest is and can prepare the table and food accordingly. For example, if the guest has a gluten allergy, the hotel can order food from a licensed gluten-free production facility. There will also be no need for the guest to mention any preferences. All the information around the guest would be already collected from Blockchain to ensure guest satisfaction.

Well-known companies have already started to use blockchain technology for a few years and have already shown successful results. TUI Group, has integrated Blockchain in the tourism sector by moving all its contracts to a private blockchain. It maintains records of hotel bed inventories in real-time. The company has also collaborated with Winding Tree on a public decentralized travel ecosystem to distribute travel inventory. The company seeks to eliminate the expensive middlemen in the tourism sector and increase the value for the guest. Another well-known company, Webjet, has recently launched its Blockchain-based platform Rezchain. The company developed, in collaboration with Microsoft, an error-free hotel bookings platform. Some European airlines, Air France-KLM and Lufthansa have explored Blockchain in the travel industry. Trippki, a project devoted to Blockchain in the tourism sector, aims loyalty program using Blockchain.

3.7.6 BLOCKCHAIN FOR HOTEL RESERVATIONS

The advantage of Blockchain in the hotel industry is that it can connect possible customers with a ray of different types of hotels without charging any commission to the hotel. It will not only be beneficial for the hotel, but the consumer will also save money in this way. Customers can pay with cryptocurrencies. Blockchain can be beneficial to get reliable reviews. Fake or unfair hotel reviews are a factor for many hotel bookings and create a lot of issues and unfair results for the hotel and the customer. Hotels will no longer pay for reviews or delete any negative reviews but rather make sure that the guest's guests' requirements are met before a negative review is posted on a public site and leave a permanent scar on the hotel's reputation. Traditional hotel booking sites are often the target for fake reviews, which can be misleading when booking a room. It is very difficult for a customer to differentiate between real and fake reviews and stay in a place that does not meet the expected standards. It will enhance Blockchain the guest experience, but it can also be beneficial for a hotel to recognize the type and history of guests that want to stay at a hotel. If a guest has a long history of bad behavior, the hotel can act accordingly before the guest arrives.

3.8 VALUE OF BLOCKCHAIN IN THE HOSPITALITY INDUSTRY

The hospitality industry is one of the forerunners in adapting to the latest trends and technology. A similar approach has recently been spotted after the emergence of Blockchain Technology. This is a potential and secure medium for document authenticating and verification. The technology is also comprehended as a great fix for the frauds and forgery in business contracts, operating agreements, and management contracts. Atlantis the Palm, Dubai, has witnessed secure technology in its payment system onsite. It will enable the guest to be charged for all their transactions at a single secure digital "wallet" that is "always-on" viewable by all parties. Such a system will provide instant access to the customer understanding after every transaction on the cloud-based system. The knowledge would be doubled and tripled with the integration of third-party entries around the city, country, and beyond. This will encapsulate all the priorities of sale, loyalty, touchpoints, etc., through the next generation of CRM (customer relationship management), which has initiated its steps through service agreements. This will normally include a complex agreement involving lawyers, third parties, and hefty fees to "validate" a specific action. Blockchain can reduce the middleman and the possibility of fraud and altering documents. Blockchain codifies documents permanently and becomes a legal document in a court case.

3.9 ELIMINATING NON-FINANCIAL FRAUD IN THE HOSPITALITY INDUSTRY

The most important contribution the hospitality industry can make to prevent or minimize non-financial crimes is getting educated in Blockchain and emerging yourself in technology. HR departments should recruit employees with a wide selection of skills to be efficient in Blockchain systems. Skills that may be considered important include economics, distributed systems, and knowledge about game theory [19]. Blockchain is in the developing stage. It will take time before the value of this technology will be recognized; however, keeping up with the latest development will place the industry ahead of fraudsters and criminals. Multiple blockchain ecosystems take advantage of the underlying capabilities of blockchain platforms, and the employer can take the time to understand Blockchain in a greater depth.

3.10 BLOCKCHAIN AND THE LATEST PANDEMIC

Like all technology investments, pandemics commonly affect new technology in different ways. The weaknesses of global supply chains were highlighted during the Pandemic, forcing companies to have a re-look at their supply chain [20]. Many industries (financial and non-financial) have seen solutions and possibilities of what Blockchain can contribute to supply and demand [21, 22]. Blockchain solutions can increase efficiency and reduce waste, including better management.

3.11 CONCLUSION

It seems that Blockchain has a bright future and could become eventually one of the most innovative ideas of the past decade. Incorporating blockchain technology in a company can create a cleaner environment. Long-term net profit on yearly financial statements can increase through the implementation of Blockchain by reducing unnecessary costs created by fraudulent information. Blockchain technology will help prevent documentation fraud in education and increase the value of qualifications that seem currently decrease because of fraudulent activities. Through

Blockchain record-keeping processes, authorization, and recordings of sensitive data will enhance the quality control of every department.

Employees' roles and skillsets may change as new blockchain-based techniques and procedures emerge in the hospitality industry. Standardization and transparency in reporting and accounting will be enhanced, enabling more efficient data extraction and analysis for future projects. When looking at the technological possibilities of Blockchain within different domains in the hospitality sector, the possibilities seem endless. It could be an exciting prospect for reducing communication error, improving coordination between two different countries, and reducing the risk of any other cybersecurity concerns.

3.12 SIGNIFICANT IMPACT OF RESULT AND FUTURE STUDY

The characteristics of Blockchain can resolve many non-financial frauds in the hospitality industry. Studying the possibilities of Blockchain in non-financial fraud cases that are not always visible in a company will open new ideas on how to combat non-financial crime in this industry. It will benefit the hospitality industry and increase the accreditation of universities and the value of degrees for employees who studied and earned the qualification through hard work. The chapter will lead to new possibilities for implementing Blockchain in the Hospitality and Tourism industry and would give a platform to many enthusiasts to develop a unique solution for Blockchain in a smarter tomorrow.

The careful planning of the technology in its infancy stage itself would provide the Tourism and hospitality industry with a chance to improve the quality by ensuring security and enhancing guest satisfaction and profitability. Verification is dependent on a consensus, which can slow down the system. As the hospitality sector starts to see the value of Blockchain and its early adaptation, it would eventually benefit the stakeholders. Blockchain technology is constantly evolving, and it can be expected that integration and interoperability standards will emerge system matures soon. More brand names adopt blockchain technology. Larger Blockchain networks will be established. Blockchain technology will eventually be recognized as a game-changer in hospitality and tourism. It will potentially reduce middlemen involvement, decrease security risks, and decrease fraudulent activities. Further studies can be conducted on how the transactions validation process can be improved and computing power can be minimized to reduce the overall cost of a blockchain system.

KEYWORDS

- **artificial intelligence**
- **blockchain**
- **employer credential fraud**
- **hospitality industry**
- **non-financial fraud**
- **tourism industry**
- **transformational change**

REFERENCES

1. Jouve Press Release, (2018). *What Can Be Done to Tackle Document Fraud?* Retrieved from Jouve: https://www.jouve.com/ (accessed on 07 September 2023).
2. Mohammed, U. S., Shukla, V. K., Sharma, R., & Verma, A., (2022). Next step to the future of restaurants through artificial intelligence and facial recognition. In: Tavares, J. M. R. S., Dutta, P., Dutta, S., & Samanta, D., (eds.), *Cyber Intelligence and Information Retrieval. Lecture Notes in Networks and Systems* (Vol 291). Springer, Singapore. https://doi.org/10.1007/978-981-16-4284-5_9 .
3. Akshay, (2019). *Why Blockchain Needs Our Help in its Fight to Survival.* Retrieved from Medium Dail Digest: https://medium.com (accessed on 07 September 2023).
4. IDC Press Release, (2020). *Global Blockchain Spending to Reach 4.1 Billion in 2020.* Retrieved from Unlock: https://www.unlock-bc.com/ (accessed on 07 September 2023).
5. Hagiu, A., (2013). Strategic decisions for multisided platforms. *MIT Sloane Management Review*.
6. WTTCO, (2019). *Geneva: World Travel and Tourism Council.* Retrieved from World Travel and Tourism Council. https://wttc.org/ (accessed on 07 September 2023).
7. IATA, (2020). *Fraud in the Airline Industry Why Carriers Need to Think of Themselves as Crimefighters.* Retrieved from Airlines: https://www.iata.org/contentassets/8a1d401955164c868258e7875edd5d5a/iata_whitepaper_fraud_july2020_digital_en.pdf (accessed on 07 September 2023).
8. Paul, A. T. D., (2011). Educational imposters and fake degrees. *Research in Social Stratification and Mobility*, 57–69.

9. Baechler, S., (2020). Document fraud: Will your identity be secure in the twenty-first century? *European Journal on Criminal Policy and Research, 26*, 379–398.
10. Allen, E. J. B., (2012). *Degree Mills: The Billion-Dollar Industry That Has Sold Over a Million Fake Diplomas.* New York: Prometheus Books.
11. Devender, M. V., (2020). *Coronavirus and Credential Evaluation: Challenges and Opportunities.* Newsletter. Retrieved from The Association for International Credential Evaluation Professionals (TAICEP): https://www.taicep.org/taiceporgwp (accessed on 07 September 2023).
12. Batalova, J., (2019). *Refugee and Asylum Seeker Populations by Country of Origin and Destination, 2000–19.* Retrieved from MPI (Migration Policy Institute): https://www.migrationpolicy.org/ (accessed on 07 September 2023).
13. UN Refugee Agency, (2020). *Figs at Glance.* Retrieved from: UN Refugee Agency: https://www.unhcr.org/ (accessed on 07 September 2023).
14. United Nations, (2015). *Envision2030: 17 Goals to Transform the World for Persons with Disabilities.* Retrieved from United Nations- Department of Economic and Social Affairs: http://www.un.org (accessed on 07 September 2023).
15. Suzanne, G., (2017). *A Recipe for Continuous Improvement Success.* Retrieved from: Food Processing: https://www.fponthenet.net (accessed on 07 September 2023).
16. Barbara, W. C. M., (2003). *Leaning into Six Sigma: A Parable of the Journey to Six Sigma and a Lean Enterprise.* New York: The McGraw-Hill Companies.
17. Keller, P., (2011). *Six Sigma Demystified.* Arizona: Mc Graw Hill.
18. Loyyal, (2017). *Loyyal Collaborates with Smart Dubai in Second Phase of Dubai Future Accelrators Program.* Retrieved from Loyyal: https://medium.com/ (accessed on 07 September 2023).
19. Pate, D., (2020). *The Skills Companies Need Most in 2020—And How to Learn Them.* Retrieved from LinkedIn Learning Block: https://www.linkedin.com/business/learning/blog/top-skills-and-courses/the-skills-companies-need-most-in-2020and-how-to-learn-them (accessed on 07 September 2023).
20. Wanganoo, L., Prasad, P. B., Tripathi, R., & Kumar, S. V., (2021). Harnessing smart integration: Blockchain-enabled B2C reverse supply chain. In: *2021 International Conference on Computational Intelligence and Knowledge Economy (ICCIKE)* (pp. 261–266). doi: 10.1109/ICCIKE51210.2021.9410677.
21. Gupta, R., Shukla, V. K., Rao, S. S., Anwar, S., Sharma, P., & Bathla, R., (2020). Enhancing privacy through "smart contract" using blockchain-based dynamic access control. In: *2020 International Conference on Computation, Automation and Knowledge Management (ICCAKM)* (pp. 338–343). doi: 10.1109/ICCAKM46823.2020.9051521.
22. Anwar, S., Shukla, V. K., Rao, S. S., Sharma, B. K., & Sharma, P., (2019). Framework for financial auditing process through blockchain technology, using identity-based cryptography. In: *2019 Sixth HCT Information Technology Trends (ITT)* (pp. 99–103). doi: 10.1109/ITT48889.2019.9075120.

CHAPTER 4

Artificial Intelligence and Invigoration of the Hospitality Industry: An Analysis

GAGANDEEP KAUR and PRASHANT CHAUHAN

School of Law, University of Petroleum and Energy Studies, Dehradun, Uttarakhand, India

ABSTRACT

The technological age has ushered in a new dimension in many fields. No field, be it technical or not, has remained untouched by its impact. The hospitality industry has progressed towards a new age of transformation with artificial intelligence (AI). AI is being utilized in the sector in several manners. AI-powered robots have replaced human attendants, and AI-based applications manage the booking process. Considering the requirement of the hospitality industry, the use of AI has become inevitable. The growing demand for personalized services has forced the industry to adopt technological means to redress the issue. The use of technology has reduced human involvement, making the entire mechanism technology-based. This chapter examines AI's role in the hospitality field, with emerging trends in the said industry. Two research questions are discussed here: firstly, how effective AI can be in the hospitality sector, and secondly, how the use of AI in the hospitality sector harms the industry. Based on the discussion, conclusions have been drawn with relevant suggestions.

4.1 INTRODUCTION

In its simple and generic sense, hospitality denotes a great approach to entertaining a visitor or gucsts. A gesture more relatable to the homely environment began to take the shape of an industry with the increasing expansion of various ancillary domains [1]. Globalization increased modes of transportation, and tourism avenues have accelerated the hospitality industry's progression. On examination of the traits of the hospitality industry, it is by a large a broader subsection of the service industry [2]. It consists of four broad branches: food and beverages (F&Bs), lodging, recreation, and tourism, and travel. If further bifurcation is made, the branches of the hospitality industry are divided into subcategories. In the F&Bs, the industry includes street food trucks, bars, and restaurants. Travel and tourism include lodgings, travel agencies, and various forms of transportation. At the same time, recreation caters to the need for entertainment, sports, and wellness-related activities [3]. All the subsectors of the hospitality industry are interconnected; they not only complement each other but, to some extent, place reliance on each other, meaning that if one sector is affected, the impact is also visible in other sectors. The hospitality industry and its related sectors are based on customer requirements. To cater to the increasing demands of the customers and the uprising of new technologies, many sub-sectors have evolved at a quick pace [4].

4.1.1 DEFINITIONS

Defining or constraining any topic or term word is a challenging task; for the feasibility of understanding the topic, defining a term is necessary. The Oxford Dictionary defines Hospitality in two ways firstly as a "friendly and generous behavior towards guests" and secondly as the mode of providing "food, drink or services that are provided by an organization for guests, customers, etc." [5] It can be said that it is a large organizational setup, catering to the needs of the customer of the hospitality industry.

4.1.2 CHARACTERISTICS

4.1.2.1 INCORPOREAL

The hospitality industry is a service-based industry, service being incorporeal and intangible, lacking the physical features to identify before they are received. Hence to enhance the quality of the product, the service staff pays special attention to providing service of personalized nature, including the etiquette, conversation, equipment quality, and a friendly environment [6]. Where the tangible relates to the product served, intangibility refers to how the product is served. The ultimate goal is customer satisfaction and expanding the customer base.

4.1.2.2 PERISHABLE NATURE

As discussed previously, the hospitality industry is embodied with intangibility. The services do not have any physical existence and hence cannot be stored for future use. Moreover, the impact is seen in a real-time experience. The services have no physical existence and hence cannot be stored for future use [7]. The use of services has a specific duration for its utilization.

4.1.2.3 WORKFORCE CENTRIC

Though the entire hospitality industry is customer-oriented, the dependability of the workforce plays a pivotal role in creating a conducive customer experience environment. The role workforce differentiates depending on the level of the organizational setup. Those eying high-end customers primarily focus on this aspect. For instance, the high star-rated hotels aim to provide services based on one-on-one mode to enhance the customer experience. With the increase in technology implementation in providing several other services, the experience based on personal touch is still considered an essential part of the hospitality industry [8]. Hence, the personal experience factor attracts a big influx of hospitality workers.

4.1.2.4 DIVERSIFICATION

Hospitality is a service-oriented industry, and every establishment tends to outperform the other through peculiar facilities and services, which is an essential factor in hospitality. This tendency developed by the establishments to cater to the customers' needs provides ample options to the consumer [9]. This type of diversification in the services and facilities facilitates the expansion of the customer base of the establishment [10].

4.1.2.5 INDUSTRY BASED ON RELATIONSHIP

In-service industry, revenue generation is an essential factor, but in the long run, the goodwill or the brand name plays an important role in the future course of the business. In developing the relationship with customers, the employees have an important role. It is evident that the employees represent the establishment; their interaction certainly impacts the organization's reputation. This relationship-building strengthens the relationship [11] with customers and maintains cordial relations with the management.

4.2 HOSPITALITY INDUSTRY

4.2.1 THE FOOTPRINTS

If we intend to trace the history of the hospitality industry, it is not feasible to trace the chronological order. The rules of hospitality were considered to be sacred in ancient Greece. Xenia was defined as "the generosity and courtesy shown to those far from home or associates of the person bestowing guest-friendship." In other words, Xenia consisted of two basic rules:

- From the perspective of the host, the host must provide all the necessities like food, drinks, a bath, and gifts, including escorting them safely to their prospective destination.
- From the perspective of the guest, the guest must not create any kind of threat or burden upon the host and must reside with courteous conduct [12].

With time, the concept of hospitality has witnessed several changes, but the basic philosophy of hospitality has remained intact. The basic objective was to make provide a conducive environment for the guest.

4.2.2 EMERGING TRENDS

Every industry has witnessed the impact of the progression of time. The same has been with the hospitality industry. The progression has resulted in various trends. Some of them are discussed briefly as in subsections.

4.2.2.1 DIGITAL TRANSFORMATIONS

With technology creating an impact all around, the hospitality industry has also not remained untouched. The management of the services provided by the hospitality industry through applications has allowed us to regulate the cycle of the guest and record their experiences [13]. Since the pandemic COVID-19, the momentum has gained towards making the services contactless and digital [14]. The services offered traditionally are being converted with thc hclp of tcchnology, i.c., voice-based applications, biometrics entry and exit, electronic payment, etc.

4.2.2.2 PERSONALIZED SERVICES

With the growing number of avenues, the customer demands more individual-centric services. The establishments have begun to tend to entertain the guest with a personal touch. Applications like Mailchimp and Zoho have revolutionized marketing through electronic means, addressing the masses with audience-oriented communication. The application that was being used for ordinary communication, in modern times, the data so collected through these means are applied in providing suggestions for future services. Technology facilitates one-to-one interaction between the guest and the establishment by utilizing big data through CRM and CEM. Customer services are being assisted by AI-based applications in booking and responding to customers' queries.

4.2.2.3 SOLO PURSUITS

A new trend in traveling has emerged where the individual tends to explore various ventures and destinations solo, experiencing self-realization. This trend has made the establishments create a more conducive atmosphere, more informal, and suits the individual. The concept of homestay has emerged as a mode of bridging the gaps between the guest and the locals.

4.2.2.4 SWITCHING TO VIRTUAL REALITY (VR)

The establishments involved in the hospitality industry have shifted their focus on the visually appealing orientation. The trends like a virtual tour allow the consumers to experience reality virtually through the digital environment. The digital experience of the ambience of the hotel or restaurant provides a first-hand experience to the consumer in making their choice [15]. The material made available in a virtual manner can reach out to the masses without the need for additional assisting devices like VR headsets. The previous virtual experience becomes handy for the guest to gather additional information through smartphones.

4.2.2.5 REMOTE WORK STATIONS

Within pandemic COVID-19, the trend of work from home gained momentum. By the time the restrictions were relaxed, the hospitality industry had explored a new opportunity to provide leisure and work together. With the shift in the global health crisis, the big companies have adopted the working remotely policy, which is more flexible and hybrid. As per the reports, the number of persons working remotely increased twice in 2021. Hence, it provides a double-sided opportunity for the customer to enjoy leisure and work in one place. The hospitality industry has encashed on the trend providing facilities like Wi-Fi, etc., to accommodate the need of the customer as per the changing environment.

4.2.2.6 DRIFT TOWARDS SUSTAINABILITY

The issue of sustainability has gained momentum across the globe, considering the environmental issues. The interdependence of other industrial sectors has created a chain. Most products used in the hospitality and allied industries owe their origin to plastic. The use of plastic not only endangers the current environment it also poses a threat to future generations. The hospitality industry has seen the paradigm shift from plastic to biodegradable material. The top management has become inclined toward eco-friendly products, including the shift from vegetarian to vegan [16].

4.2.2.7 HEALTH TOURISM

The modern lifestyle of the individual has adversely impacted their health. Nowadays the person is becoming more and more concerned about his/her health. This diversion in health trends has given rise to an opportunity for the hospitality industry in the form of health tourism. There are two facets of health tourism: first, related to the existing medical issues that require treatment; second, health tourism assists in physical and mental health and showcases a preventive approach by promoting the concept of self-care [17]. The pandemic has promoted the wellness industry, which is becoming a trillion-dollar industry. These wellness centers provide personalized treatments by experts under one roof.

4.2.3 *ROLE OF THE HOSPITALITY INDUSTRY*

The hospitality industry plays an important role in the economy. As per the data available, the estimated contribution to the global GDP of travel and tourism was around 6.1%. For the year 2021, they depict an increase compared to 2020. Approximately 5.81 trillion dollars was contributed to the GDP by the travel and tourism industry [18]. The contribution to the GDP by travel and tourism is depicted in Figure 4.1. Growth is shown in terms of percent.

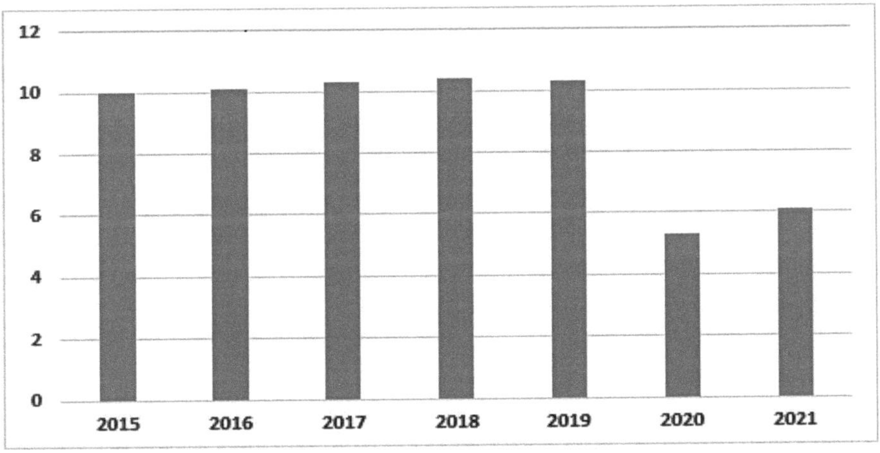

FIGURE 4.1 Share of GDP from travel and tourism.

Source: Travel and Tourism: Share of Global GDP (2021, 2022).

With the expansion of travel infrastructure, the international arrival witnessed an increase to 1.3 billion from 900 million in just the last 10 years. This resulted in the creation of employment opportunities globally, which boosted the hotel industry. Although the pandemic COVID-19 adversely affects the travel sector. The promising annual growth in the travel sector witnessed a drop to 73% for the year 2020 and around 73% in 2021, respectively [19]. This unprecedented decline resulted from the travel restrictions and the countries closing their border to stop the spread of the virus. Year 2020 is one of the worst years in the history of travel and tourism, where the international travel recorded was below 1 billion, resulting in a revenue generation of just around 1.1 trillion US dollars. According to World Trade Organization (UNWTO), around 74% decline has been witnessed from the level that was 30 years ago [19]. The UNWTO, in its prediction for the year 2021–2024, stated that it would take around two and half years to regain the status of the year 2019 concerning intentional travel. Although the results for the year 2022 have been positive.

WTTC indicated the hospitality industry as a major component in terms of value creation around the globe. According to the council, the global contribution of the industry has been just below 6.4% before the pandemic, with around 8.6 trillion US dollar revenue contribution to the world economy.

The hospitality industry is one of the viable sources of employment generation, accounting for one out of the 10 employment opportunities. Around 330 million people were earning their livelihood from this sector. The figure showed a decline of around 1% compared to pre-pandemic data. The year 2020 showcased around 58 million jobs in the sector, increasing the percentage up to 21.5%. The rise in revenue is a constituting factor in creating employment for the population at local destinations; the hotel owners or the companies also derive benefits from the rise in revenue. Before the pandemic COVID-19, approximately 173 million people were earning their livelihood from the motel and hotel industry [20]. It is estimated that the travel and tourism growth rate is around 15.1%. The global hospitality market is expected to grow from 3,952.87 billion dollars in 2021 to 4,584.42 billion dollars in 2022. With travel restrictions waived, the hospitality sector is expected to witness a rise in revenue to 6,715.27 billion dollars for the year 2026, with the CGAR at 10.2% [21].

4.3 ROLE OF AI IN THE HOSPITALITY INDUSTRY: AN ANALYSIS

Technology has brought changes in every aspect of human life; hospitality has not been an exception. Futuristic technology like AI has brought several changes in the hospitality sector. It is often seen as an apprehension that technology like AI might adversely impact the human workforce. However, AI has been paving the way for hospitality to be more guest-centric. Many leading brands across the globe are placing reliance upon AI-powered systems and robotic servers to provide more enhanced services. Some of the ways AI is transforming hospitality are being discussed as in subsections.

4.3.1 AI-INFUSED ROBOTIC ATTENDANTS

Several establishments have realized that client insight is the key to improving the brand's value. In recent times AI-infused concierges' services have gained momentum. Hilton concierge is not only revolutionizing and enhancing the guest experience. Similarly, Connie, a personable robot powered by AI based on IBM Watson, provides more personalized service to the customer. Japan has been using such AI-powered robots at reception to handle the customers [22]. These AI-powered devices can interact with customers, take up their questions, and answer appropriately. Not only these devices can recognize speech for communication; they are programmed in a manner to adapt and learn from the experience, to improvise their responses to interactions.

4.3.2 AI INFUSE VIRTUAL ASSISTANT

The Chatbots developed with AI have depicted impressive results in their operation in hospitality. These chatbots are specifically programmed and trained to address task-based user needs. These are the new face of customer service. Application-based upon AI has shown exceptionally compelling results, specifically concerning online matters. These chatbots are available 24/7 to assist the customer, breaking the barriers of time restrictions. The establishments can reach out to the customers through these chatbots, helping in a great user experience. Apart from interacting with customers, these chatbots can take up consumer complaints, place orders, manage reservations, etc. [23]. The most beneficial aspect of AI-based devices is the freedom of accessibility. With a smartphone, these can be managed or controlled, enhancing user experience.

4.3.3 BOOKING PROCESS

The manual booking process is more time-consuming and a painstaking process. With most of the activities being conducted through online platforms, managing them can be a challenge. The AI-based booking process can provide a more convenient and reliable booking process. AI-based applications can handle and manage a large amount of data, which is useful in analyzing the availability of empty booking slots and providing viable options to the customer. This can reduce friction and enhance the sale potential through an improvised booking experience for the customer.

4.3.4 AI-BASED REVIEW

The technology through online platforms allows the user to share their experience about the service availed. The customer feedback not only allows the establishment to improve their services based on the feedback but also helps prospective customers decide the services be picked. Instead of visiting every review-providing site, the customer can seek assistance from an AI-powered application to segregate the review or feedback as per the user's choice. This application can help provide real-time information to the user.

4.3.5 PRICING PREDICTIONS AND ANALYSIS

Although the hospitality sector is a running industry, certain seasons witness a massive influx of customers. To attract customers during peak seasons, the establishments provide lucrative offers; in this, the AI base applications can be beneficial in monitoring the demand and determining the pricing depending upon the market. Predictive analysis helps enhance earning potential by examining multiple factors.

4.4 CHALLENGES IN USING AI IN THE HOSPITALITY INDUSTRY

As per the above discussion, there are two major issues that a posing challenge in the application of AI in the hospitality industry. They are discussed briefly in subsections.

4.4.1 UNDERDEVELOPED

As AI is still developing, it is a significant challenge in its implementation. As the consequences of its potential are not yet being realized. AI has shown tremendous progress and potential despite being at an infant stage. The vice president, in an interview, stated that "AI is currently very, very stupid... It's not something we could press to do general-purpose reasoning involving things like analogies or creative thinking or jumping outside the box." The infancy might affect the ability to analyze the data and make a reasonable decision. As AI act upon the data fed through an algorithm, and the data is fed by a human, there is the possibility that the data collected by the creator might be biased and impact the acts of the machine.

4.4.2 ISSUE OF DATA PRIVACY

This is the most striking factor in the use of AI. As AI is based upon data, the apprehensions might be whether the data collected for the applications is safe or not. The hospitality industry is completely based upon human interaction; the use of AI applications will ultimately collect the individual's private data. The data submitted may be deliberate or unintentional, but the machine will store and analyze it. How will it be ensured that the data is secured? The risk of data violation is always present. A study showed that 9 out of 10 internet users were concerned about their data protection.

4.5 CONCLUSION AND SUGGESTIONS

4.5.1 CONCLUSION

At the current time, the impact of technology cannot be evaded. The hospitality sector is a service-based industry. The service industry always looks to improve its service for a better customer experience. The value of the industry is dependent upon customer feedback and response. AI is undoubtedly assisting the enhancement of the hospitality industry in the ways discussed above. However, the issue of data protection is a challenging task that requires immediate redressal. Securing the private formation is not only necessary for the individuals' rights but also to maintain the credibility of the hospitality industries.

4.5.2 SUGGESTIONS

Effective mechanisms are needed for data protection in the hospitality industry as the entire industry is based upon individual interaction. Additionally, the trust factor is considered one of the most prominent components in the hospitality industry; therefore, for better customer experience, a robust policy framework should be incorporated with the assistance of AI.

KEYWORDS

- **artificial intelligence**
- **data privacy**
- **health tourism**
- **hospitality industry**
- **robotic attendants**
- **travel and tourism**
- **virtual assistant**

REFERENCES

1. King, C. A., (1995). What is hospitality? *International Journal of Hospitality Management, 14*(3, 4), 219–234. https://doi.org/10.1016/0278-4319(95)00045-3.
2. Ali, B. J., Gardi, B., Othman, B. J., Ahmed, S. A., Ismael, N. B., Hamza, P. A., Aziz, H. M., et al., (2021). Hotel service quality: The impact of service quality on customer satisfaction in hospitality. *International Journal of Engineering, Business and Management, 5*(3), 14–28. https://doi.org/10.22161/IJEBM.5.3.2.
3. Bansal, S. P., Gautam, P. K., & Kumar, S. A., (n.d.). *Tourism & Hospitality Front Office Operations & Management Hospitality Products and Services Principal Investigator Co-Principal Investigator Content Reviewer.* Module-04 "Earlier Hospitality Types and Concepts."

4. Wong, O. M. A., Dean, A. M., & White, C. J., (1999). Analyzing service quality in the hospitality industry. *Managing Service Quality: An International Journal, 9*(2), 136–143. https://doi.org/10.1108/09604529910257920.
5. Definition of hospitality, (n.d.). In: Oxford Advanced Learner's Dictionary. Retrieved from: https://www.oxfordlearnersdictionaries.com/definition/english/hospitality?q=hospitality (accessed on 07 September 2023).
6. Păuna, D., & Filip, M. L., (2015). Assessing goodwill in hotel business case study: Milcofil SA. Vrancea. *EIRP Proceedings*. https://papers.ssrn.com/sol3/papers.cfm?abstract_id=2595260 (accessed on 07 September 2023).
7. Croes, R., & Semrad, K. J., (2012). Discounting works in the hotel industry: A structural approach to understanding why. *Tourism Economics, 18*(4), 769–779. https://doi.org/10.5367/TE.2012.0138.
8. Baum, T., (2006). Skills and training for the hospitality sector: A review of issues. *Journal of Vocational Education & Training, 54*(3), 343–364. https://doi.org/10.1080/13636820200200204.
9. Zheng (Jerry), C., & Tsai, H., (2019). Diversification and performance in the hotel industry: Do board size and family representation matter? *International Journal of Contemporary Hospitality Management, 31*(8), 3306–3324. https://doi.org/10.1108/IJCH.
10. Kang, K. H., & Lee, S., (2014). The moderating role of brand diversification on the relationship between geographic diversification and firm performance in the US lodging industry. *International Journal of Hospitality Management, 38*, 106–117. https://doi.org/10.1016/J.IJHM.2013.01.001.
11. Sota, S., Chaudhry, H., & Srivastava, M. K., (2019). Customer relationship management research in hospitality industry: A review and classification. *Journal of Hospitality Marketing & Management, 29*(1), 39–64. https://doi.org/10.1080/19368623.2019.1595255.
12. *Xenia*, (n.d.). Britannica. Retrieved from: https://www.britannica.com/topic/xenia-sociology (accessed on 07 September 2023).
13. Cvetanka, R., & Dimitrov, N., (2019). Digitalization in the hospitality industry - trends that might shape the next stay of guests - UGD Academic Repository. *International Journal of Information, Business and Management, 11*(3). http://eprints.ugd.edu.mk/21976/ (accessed on 07 September 2023).
14. Morrone, D., Raimo, N., Tarulli, A., & Vitolla, F., (2021). Digitalization in the hospitality industry: Motivations, effects and role of COVID-19. *International Journal of Digital Culture and Electronic Tourism, 3*(3, 4), 257. https://doi.org/10.1504/IJDCET.2021.116475.
15. Nayyar, A., Mahapatra, B., Le, D., & Suseendran, G., (2018). Virtual reality (VR) & augmented reality (AR) technologies for tourism and hospitality industry. *International Journal of Engineering & Technology, 7*(2), 156–160. www.sciencepubco.com/index.php/IJET (accessed on 07 September 2023).
16. Dos Santos, R. A., Méxas, M. P., & Meiriño, M. J., (2017). Sustainability and hotel business: Criteria for holistic, integrated, and participative development. *Journal of Cleaner Production, 142*, 217–224. https://doi.org/10.1016/J.JCLEPRO.2016.04.098.

17. Mueller, H., & Kaufmann, E. L., (2016). Wellness tourism: Market analysis of a special health tourism segment and implications for the hotel industry. *Journal of Vacation Marketing*, *7*(1), 5–17. https://doi.org/10.1177/135676670100700101.
18. *Travel and Tourism: Share of Global GDP 2021*, (2022). Statista. https://www.statista.com/statistics/1099933/travel-and-tourism-share-of-gdp/ (accessed on 07 September 2023).
19. *The Economic Contribution of Tourism and the Impact of COVID-19*, (2021 World Tourism Organization (UNWTO). https://doi.org/10.18111/9789284423200.
20. Jones, P., & Comfort, D., (2020). The COVID-19 crisis and sustainability in the hospitality industry. *International Journal of Contemporary Hospitality Management*, *32*(10), 3037–3050. https://doi.org/10.1108/IJCHM-04-2020-0357/FULL/PDF (accessed on 07 September 2023).
21. *Global Hospitality Market Report 2021: Market is Expected*, (n.d.). Retrieved from: https://www.globenewswire.com/news-release/2021/06/08/2243615/28124/en/Global-Hospitality-Market-Report-2021-Market-is-Expected-to-Reach-5297-78-Billion-in-2025-Forecast-to-2030.html (accessed on 07 September 2023).
22. Drexler, N., & Beckman, L. V., (2019). For better or for worse: Shaping the hospitality industry through robotics and artificial intelligence. *Research in Hospitality Management*, *9*(2), 117–120. https://doi.org/10.1080/22243534.2019.1689701.
23. Berezina, K., Ciftci, O., & Cobanoglu, C., (2019). Robots, artificial intelligence, and service automation in restaurants. *Robots, Artificial Intelligence and Service Automation in Travel, Tourism and Hospitality*, 185–219. https://doi.org/10.1108/978-1-78756-687-320191010/FULL/XML.

CHAPTER 5

Applications of Robotics in the Hospitality Industry

VENKAT SAI KARANAM,[1] VINOD KUMAR SHUKLA,[1] MOHAMMAD NADEEM KHALID,[2] and SOUMI DUTTA[3]

[1]*Department of Engineering and Architecture, Amity University, Dubai, UAE*

[2]*Institute of Applied Technology, Abu Dhabi, UAE*

[3]*Institute of Engineering and Management (IEM), West Bengal, India*

ABSTRACT

Hospitality Industry is a service-oriented industry aligned to serve guests, tourists, and customers with their altering requirements, service standards, expectations, and technological influences. In recent years, the technology has been immersive in the varying convenience and support apprehensions of the client, which the service providers have outmoded. The implantation of techno-based service tools in the industry is not the latest trend, nor the amusement of its features; however, the ground-breaking innovation and its applications in the hospitality industry has been potentially effective in raising the service standard and economically efficient. Multiple industries were using the robots in several fortes. However, the hospitality industry brought a real-life experience and a chance to rub shoulders with the robots to an ordinary man. Hospitality industries strive to delight customers by fulfilling their expectations, which invites multiple repeated actions with slight variations from time to time. The proper application of Robotics has transformed the industry and its processes, making it more accessible, convenient, and economically efficient. The chapter highlights practical implementations and their relative benefits and impediments.

5.1 INTRODUCTION

Robots have changed everything from self-driving cars to cleaning services. Robots have also changed things for the hospitality industry where they use robots to provide essential services like payment services, cleaning, customer care and even room service. Various countries have started to adopt robotic hospitality, and integrating robots has become more accessible for the customers (like translation). Robots are becoming more of a common thing in various hospitality services. Many of the leading Hotel chains are aware of what robots are capable of as they are cheap and inexpensive to maintain, but since the technology is not that good right now, some hotels have fired robots since they cost the same as humans. However, most hotels use them for concierge services. These robots can also give basic information such as hotel services, check-in/check-out time, nearby tourist spots and points of interest.

The global industrial robots market has predicted extensive growth in the coming years. The market is expecting to cross 70,564 million dollars of revenue by 2029 and would get fast-paced after the year 2025 extensively as expected in 2022 (Figure 5.1) [1]. The global robotics technology market was valued at $62.75 billion in 2019 and is projected to reach $189.36 billion by 2027, and growing at a CAGR of 13.5% from 2020 to 2027 [2]. According to the Association for Advancing Automation (A3), the North American automation market had set new records very recently. For the first nine months of 2017, 27,294 orders of robots, valued at approximately USD 1.473 billion, were sold in North America [1].

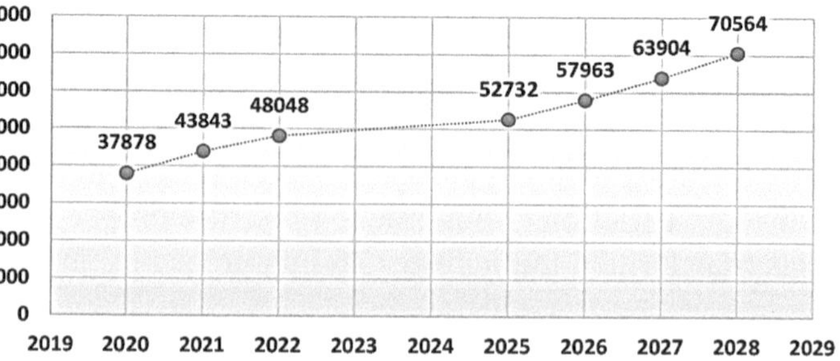

FIGURE 5.1 Global robot market (revenue in millions of dollars).

Robotics is a domain where the study of making and building machines can perform assignments that people customarily perform. These robots are coordinated by computers and may carry out basic to complex tasks.

Hospitality is a vital Industry encapsulating multiple small industries under its wings, such as tourism, transportation, restaurants, accommodation, service, airlines, casinos, leisure, gaming, events, food, beverages, etc. This provides immense opportunities for the robot's implementation in several processes and touchpoints. It leverages the satisfaction of guests and operational efficiency through always maintaining the service standards and for as many numbers of times. This reduces the chance of reduction or unavailability of services at odd hours in round-the-clock operations of the service sector dealing with accurate real-time data and varied requirements of the client. Robots assisting the flow of operations have been practical. It has been not due to the gimmick of eye-catching robotic movement, but the sensible implementation of AI has resulted in smooth operations and accurate services. Restaurants have replaced physical menus with QR codes, housekeeping with vacuum robots, tablets have taken the room control in hotels, kiosks replace receptions, bots do feedbacks, and cloud-based food ordering mechanisms evolved the industry [3]. Smart rooms and contactless service has proved the solution in the pandemic era with its keyless entry, wireless device charging, room and housekeeping services with robotic vacuum and bots [4].

5.2 ROBOTICS AND ARTIFICIAL INTELLIGENCE (AI)

Robotics is a stream of technology that deals with programmable machines equipped with sensors and actuators for interactive and autonomous, actionable devices named robots. Artificial intelligence (AI) is a computer science branch that outsets algorithms for problem-solving, logical reasoning, learning, and perception [5]. Robots and AI are different technologies, while their combination Robots implement various sensors. Internet with AI can give us outstanding results in the hospitality industry. Communication with customers can be eased using translation, offering recommendations and suggestions as per the customer's previous experiences. This can also be modified to fit the customer and provide the most service. Humans can be limiting, but data can be analyzed quicker and processed with robots and AI. Robots have become essential as information holds the most power and service. Most of the travelers stick around to countries where they find similar language. Even if they attempt to try a new culture, it will remain untouched due to the language barrier alongside. If they hire a translator, they need to depend on his availability. The possibilities of AI and robots may turn this into an opportunity to let humans understand any language of the world, unlocking the restricted tourists in the countries due to language barriers [6].

The components of AI and robotics (Figure 5.2) are the amalgamation of several giants technologies combined in a meaningful way to solve multiple sets of complex actions with the help of actuators, sensors, and algorithms in an electric supplied set of augmented AI through cloud computing and machine learning.

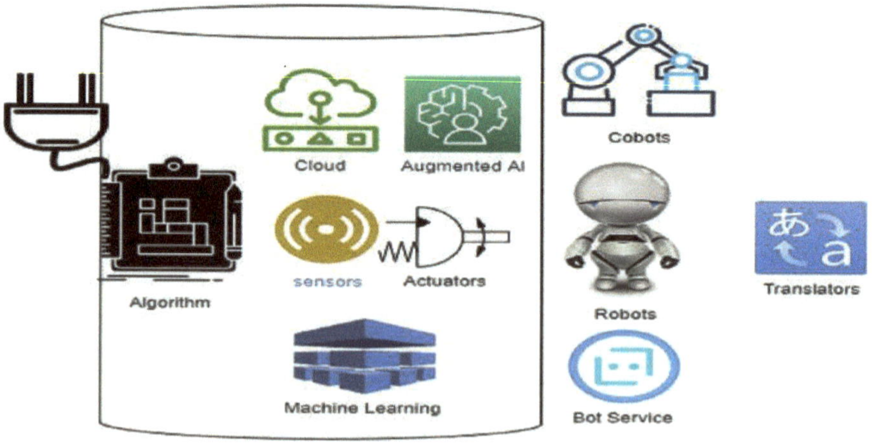

FIGURE 5.2 Components of AI and robotics.

5.3 IMPLEMENTATION OF THE ROBOTS IN INDUSTRY

Robots have been used for delivering basic amenities such as towels, pillows, toiletries, and coffee. These are also helping with the current pandemic (COVID-19), which requires maintaining a safe distance. Robots are being used in immigration services, decreasing the queue line and speeding up the process. Airport check – INS has also been integrated with AI, decreasing queue lines. Robots offer a contactless experience, a great option that limits the contact spots between guests and staff, increasing safety and health throughout the facility. During the COVID-19 pandemic, robots can do various tasks, and it also helps with safety measures. Robots cannot get ill or tired; they can take guests' temperature, transport luggage and provide various services without any human contact [7].

Robots can work for various hours before recharge, which means they can work for extended hours and do not require any salary or monetary payment. They only need a break for maintenance, software updates, and recharge. They can even help boost the efficiency and productivity of various hospitality services like checking in guests, managing guest requests, and satisfying requests that can be done without human contact. Robots can help with hygiene, sanitation, and housework services essential every time a new guest arrives. Cleaning services like vacuum and moping services can also be taken over by these robots, which saves time and increases the efficiency of the hotel staff. Robots can help create a better experience for regular guests as robots can store data and preferences about every guest, which can help the hotel/staff to help make the stay for the guest more convenient and comfortable.

5.4 ROBOTICS AND ITS IMPEDIMENTS

With these technologies, there are various setbacks, such as high installation costs; even though robotic technology is readily available for use, this technology could be expensive, and these technologies might not be available for every hotel to use. All robots require maintenance as technicians are required to maintain these technologies, and technicians require high payouts. Robots could be vulnerable to hacks and errors in their systems; an intruder could access the robot or even steal data from the hotel, which might be valuable. Errors in the robots could result in a delay in service for the guest or customer. Robots might take over these jobs, which account for about 30% of jobs in the world, which would have a negative impact, mainly due to automation [8].

Robots and AI have played an essential part in hospitality and tourism. Robots can carry out basic, easy, and complex human tasks at any time. Deploying robots could save a lot of time and money for many hotel owners, travel agencies, and even the airport. Robots can remove mistakes and errors and deliver excellent service to travelers, passengers, and guests. Hotel staff should reach guests with their name, booking requirements, last visits reviews and anticipated choices. Moreover, to make his dream visit true by leaving a note of his special, memorable day card or a free subscription to his favorite channel seems a lot when it is to be done in each shift. This will let one realize the hard work needed to make personalized service a trend in the growing hospitality industry. In such overrated service industry requirements, the implementation of Robotics and AI is not supernova effects though the mandate to utilize AI to get individual preferences.

5.5 ROBOTICS AND CUSTOMER SERVICE IN THE HOSPITALITY INDUSTRY

Robots are replacing customer service in most of the Hospitality industry. As customer service is essential for the industry, robots can deliver the best service to their customers and improve as robots can remember and recommend better things for the guests. One of the most important challenges hotels face is responding to the customer quickly. Robots can help with this as robots can process information faster and even have a huge information resource for the robot to provide the customer with. As robots can analyze the collected data, this can help robots further enhance the guest's experience.

Robots have various functions such as customer interaction to deliver drinks, customer communication, hospitality services, and safety. Artificial Intelligent Agents can take care of bookings, hotel services, customer service and complaints. Analysis AI can take what the guest/customer wants when they stay, like drinks and food. Translation Agents can take care of customers who do not speak the native language and can help with translation, which is a massive benefit as hotels always get foreign customers who do not speak the native language. There can be a lot of mechanical machines which can help with hospitality. However, robots are more sophisticated devices that can adapt and be more customizable for the hotel or the guest, such as a robotic cook who can make food depending on the guest's preference. Most robots in the Hostel or Hospitality industry are considered service robots (AI integration) since they can socially communicate with the individual. Service robots are mainly used for assistance or communication for the hotel guest to provide more services [9].

While robots have become more useable since they are easier to maintain and provide faster service, researchers are looking into more human-like AI, making it more efficient for hospitality services. These robots are being taught various aspects of a more human-like AI like mathematics, philosophy, economics, neuroscience, psychology, and linguistics (Languages). These subjects will help AI to become more and more proficient at completing various tasks. The hospitality industry is an industry that requires long working hours and low salaries for employees. This industry has integrated its service through computerized processes and AI, which help reduce and save money in customer service and CRM and create better customer preferences or recommendations. Many smart appliances are being used in hospitality, which is close to robotics. Smart appliances by Amazon, Google, and Microsoft have helped the hospitality industry reach complete computerization of services. Robotic machines such as robotic housekeeping, robotic customer service, robotic concierges, and robotic deliveries. Many famous hotels like Henna-na and Aloft Hotels have adopted a robotic hotel. Many countries have also adopted robotic tourist information centers. Robots are also being used in day-to-day jobs in the hospitality industry.

5.6 ROBOTICS INFLUENCING PROCESSES IN HOSPITALITY INDUSTRY

Current robot systems have functions crucial to tourism and hospitality, like speech recognition and autonomous planning. The usage of Autonomous technology is increasing in tourism and hospitality worldwide. One of the most common examples of Robots being implemented in hotels is at a hotel called Henn – which is in Nagasaki, Japan, and robots completely staff it from front desks to luggage transportation.

Savioke is a leading company that creates robots for hospitality services, and they have also created a robot called Relay. This robot is about one meter tall, and it is a robot that delivers items in a locked container from the front desk to the guest and vice versa. The robot has AI integration to navigate a hotel without any guidance. It uses laser technology to navigate through the hotel. The guest calls upon the front desk requesting an item, and the front desk sends Relay (the robot) to the guest. The robot navigates through the elevators and reaches the guest. Once it reaches the guest's room, it calls upon the guest, and the guest is required to authenticate that it is them. After that, they receive their items, which leaves after confirmation of items taken, including feedback of the experience. There are many robots like Relay which help with basic needs.

In-person customer service has been changed by technology companies and various hotels where they have changed by delivering an AI which can interact with humans easily. This technology can deal with simple customer service situations.

IBM is a leading company that creates various technology-based devices; they have created a robot concierge called Connie in collaboration with Hilton Hotels. This robot has gone through training in one of Hilton's hotels. It uses AI developed by IBM to interact with customers and answer their questions. It even has speech recognition technology, which also learns and adapts every time it has an interaction. It is an improving device that can help with basic customer service in hotels and other areas. The robot can also provide touristic attractions around the customer, and it can learn human behavior and adapt to different customers [10].

TravelMate is another robot used in luggage and suitcases [11]. This suitcase has a robot integrated, allowing it to follow the suitcase's owner. It uses laser technology to maneuver and has anti-collision technology. This robot removes the element of carrying or pulling around your suitcase. It is a perfect robot for those who need extra hands or grabs some snacks for their journey.

Toshiba is another company that makes robots for tourism and hostel. They created a Chihira-Aico that helps guide customers (travelers) and keep them happy with the information they need [12]. This robot assistant can do a variety of jobs, from information service to room service. This robot's key ability is that it can speak in various languages, perfect for tourists to interact with.

Amadeus has developed an experimental robot called 1A-TA, which uses AI which helps with customer service during busy hours of a Travel Agency. Travel Agencies has used this robot to get the customer's needs to be fulfilled during rush hours. This robot can find out what the customer needs and process the information for the Human Travel Agent to use this information for when the customer speaks with a human travel agent [13].

Some places where robots could be used are technologically advanced where they are effective is when it comes to online-based customer service. Online bots can help with simple questions and requests online. The bot can also virtually chat with the customer to solve more complex issues or questions. AI has been used in various platforms like social media, news platforms and online shops. It allows the customers to ask simple questions, and the customers receive an immediate answer from the bot. This improves the experience for the customer and saves time. This is something that human customer services cannot achieve.

Robots exist physically and virtual; SnatchBot is a chatbot that helps book flights or hotels online. It can create a travel booking using the SnatchBot platform. It guides the customers through a booking process and even clarifies their questions regarding their booking. This bot is available across various major platforms like Skype, Messenger, and virtual assistants [14].

Security Robots such as KnightScope security robots developed by KnightScope are being tested and used in airports. They can prevent, detect, and report any incidents on-site, that too autonomously.

This is an excellent option as it can detect concealed weapons and keep passengers (travelers) safe during their journey. It is also available in various areas other than airports [15].

Kate the robot is a robot developed by SITA Labs. This robot is an autonomous kiosk that adapts to the environment and moves around depending on the crowd [5]. This kiosk has various data types and sources, from flight details to passenger details. This robot can be implemented where additional kiosks are needed and reduce the queue time for passenger check-ins. These robots can be deployed autonomously or manually by the operator. These kiosks can provide information, check-ins, help passengers and even manage the flow of passengers by moving to a more crowded area where additional kiosks are required. The robot uses geolocation to move around and displays essential information using the airport's Wi-Fi network. The robot also uses laser and anti-collision technology to move freely without bumping or crashing into anything or anyone.

Leo the robot is another robot developed by SITA labs. This robot is an autonomous self-checkout luggage vehicle which can take care of the passenger's luggage by taking it off them and dropping the baggage at the baggage drop area. Together, these robots have helped ease the travel experience for the passengers and the operators. These robots (Leo and Kate) have helped reduce queues and save time for passengers and operators. Deploying these kinds of robots can ease the airport staff. The airport staff would save time and take care of more complex and vital tasks [17].

Nadine is a robot developed by NTU (Nanyang Technological University, Singapore), where the robot acts like a human (socially), and this robot can be used as a greeter [18]. This robot needs to interact with people to improve its functionality. This robot can even take care of elderly customers as this robot can act as a companion. This robot can interact with its surroundings, project emotions, and work for extended periods. This robot could also be used as a front-line worker for the hospitality industry as the robot can do a lot more things than all the robots combined in Henn-na Hotel in Japan. With workforce issues or saving efforts in mind, academic literature reinforces the idea that technological development in robotics, AI, and science create opportunities to enhance administrative functioning, efficiency, and excellent reliability.

Robots can operate with more accuracy and consistency than any human can do in most circumstances. However, robots can operate in harsh or hazardous areas as they can be trained to perform specific actions which Humans cannot do. Robots are evolving as valuable solutions in the hospitality industry, but robots can improve computerization (robotics) and execute a task without getting weary; it is essential to recognize there are always shortcomings like the lack of ability to adjust to conditions.

5.7 IMPACTS OF ENHANCED USAGE OF ROBOTS IN HOSPITALITY INDUSTRY

Robots are seeping through the hospitality and tourism industry; their existence is growing and continues in the industry. In future times, technicians will replace managers as they have specific technical skills to operate these machines and AI. This will maximize the efficiency, but it will also increase the benefits of using robots. If these robots are not maintained or their technology is not managed correctly, their capabilities will reduce over time. We can never rely entirely on technology, as glitches and malfunctions might occur. Therefore, there must be human involvement when there is technology involved.

Managers should either be trained or replaced by a Robotic assistance team to handle the AI and robots to have continued service. Moreover, it will improve the overall guest experience. Technology and its algorithms are unpredictable, and what it will bring for us in the future it is learning and adapting. The hospitality and tourism leaders' skills have changed how technology has evolved in this industry. The skills such as excellent service, which is being portrayed through technology, will never change as robots interact to create a friendly and enjoyable experience for the guest (traveler, customer, or client).

Travel specialists are placed to help create improved experiences available through AI and robotic assistance methods. Human skills are exceptionally critical for a successful system for the hospitality industry. Hospitality specialists should interact with the technology developers to create machines and AI to interpret outstanding people skills. In the end, hospitality and tourism specialists will remain vital elements of the hospitality industry as technology evolves in the industry. Customers may be cynical at first with the introduction of robots and freak out. However, people will slowly understand the importance of robots and how robots have evolved into useful tools such as phones, ATMs, kiosks, computers, and other gadgets. Robots cannot replace people altogether, but they are more tools for the people. Robots are tools, but they increase the efficiency of the work and save time; they can also provide a fantastic experience.

5.8 CONCLUSION

Robots and hospitality much-opposed combination has turned out a beneficial amalgamation of humans and technology, offering several front-end benefits and long-lasting business solutions. This was initially opposed and denied due to the gimmick-based techno attraction to lure guests. The evolving internet usage and technological advancement of societies have resulted in the core business, and financial profits made non-believers agree that the proper implementation of the technology will result in success. The several forms of the industries are illustrated in the chapter supports the growing trend of automation and robotics in the hospitality industry.

KEYWORDS

- **artificial intelligence**
- **geolocation**
- **hospitality**
- **robotic assistance methods**
- **robotics**
- **service satisfaction**
- **tourism industry**

REFERENCES

1. *Industrial Robotics Market Size 2021 – Latest Research Report to Share, Growth Strategy, Industry Share, COVID-19 Impact Analysis, Key Findings, and Forecast by CAGR Till 2024. MarketWatch.* https://www.bloomberg.com/press-releases/2022-05-17/demand-for-global-industrial-robotics-market-size-share-is-expected-11-8-cagr-rise-will-hit-to-usd-81-4-billion-globally-by (accessed on 07 September 2023).
2. *Industrial Robotics Market Revenue Worldwide 2018–2028.* Statista. https://www.statista.com/statistics/760190/worldwide-robotics-market-revenue/ (accessed on 07 September 2023).
3. Bob, R., & Sarah, A. *Hotel Robotics Thriving in 2020* | Hospitality Net. https://www.hospitalitynet.org/opinion/4100756.html (accessed on 07 September 2023).
4. *Smart Rooms, Mobile Check-in & Robots Room Service: Hospitality Trends to Watch Out in Year 2021*, (2021). The Economic Times. [Online]. Available: https://economictimes.indiatimes.com/magazines/travel/smart-rooms-mobile-check-in-robots-room-service-hospitality-trends-to-watch-out-in-year-2021/articleshow/79754408.cms (accessed on 07 September 2023).
5. Owen-Hill, A. *What's the Difference Between Robotics and Artificial Intelligence?* https://blog.robotiq.com/whats-the-difference-between-robotics-and-artificial-intelligence (accessed on 07 September 2023).
6. Cain, L. N., Thomas, J. H., & Alonso, Jr. M., (2019). From sci-fi to sci-fact: The state of robotics and AI in the hospitality industry. *Journal of Hospitality and Tourism Technology, 10*(4), 624–650. https://doi.org/10.1108/JHTT-07-2018-0066.
7. Chris, C., (2021). Angie Hospitality. *Is Robotics Taking Over Hospitality? Yes and No.*, PhocusWire, www.phocuswire.com/is-robotics-taking-over-hospitality (accessed on 07 September 2023).

8. Anna, L. D., (2020). *How to Use Artificial Intelligence in the Hospitality Industry*. Revfine.com, www.revfine.com/artificial-intelligence-hospitality-industry/ (accessed on 07 September 2023).
9. *8 Examples of Robots Being Used in the Hospitality Industry*. Revfine.com, https://www.revfine.com/robots-hospitality-industry/# (accessed on 07 September 2023).
10. IBM, *Hilton and IBM Pilot Connie. The World's First Watson-Enabled Hotel Concierge*. https://www.linkedin.com/pulse/hilton-ibm-pilot-connie-worlds-first-watson-enabled-hotel-gagn%C3%A9/ (accessed on 07 September 2023).
11. *Travelmate Robotics*. https://travelmaterobotics.com/ (accessed on 07 September 2023).
12. *Toshiba's Robot is Designed to be More Human-Like* – BBC News. https://www.bbc.com/news/technology-35763917 (accessed on 07 September 2023).
13. *Could Robots Become the Next Travel Retailers*. Amadeus | The leading travel technology Company. https://amadeus.com/en/insights/blog/robots-become-next-travel-retailers (accessed on 07 September 2023).
14. *SnatchBot: Free Chatbot Solutions, Intelligent Bots Service and Artificial Intelligence*. https://snatchbot.me/ (accessed on 07 September 2023).
15. Knightscope. https://www.knightscope.com/ (accessed on 07 September 2023).
16. *Kansai Airports Trial SITA Robot KATE to Slash Queues and Reduce Wait Time*. https://www.traveldailymedia.com/kansai-airports-sita-robot-check/ (accessed on 07 September 2023).
17. *SITA Develops Automatic Baggage Robot Leo*. https://www.airport-technology.com/news/newssita-develops-automatic-baggage-robot-leo-5667910/ (accessed on 07 September 2023).
18. Thalmann, N. M., Tian, L., & Yao, F., (2017). Nadine: A social robot that can localize objects and grasp them in a human way. In: Prabaharan, S. R. S., Thalmann, N. M., & Bhaaskaran, V. S. K., (eds.), *Frontiers in Electronic Technologies* (Vol. 433, pp. 1–23). Singapore: Springer Singapore. doi: 10.1007/978-981-10-4235-5_1.

CHAPTER 6

Future Employments in the Hospitality Industry

WONDA GROBBELAAR[1] and AMIT VERMA[2]

[1]*Nature Science Research and Innovation Center, Canada*

[2]*HTMi Switzerland Dubai, United Arab Emirates*

ABSTRACT

Change in the working environment is inevitable in the world we are living in today. The pandemic is driving changes to customers' behavior in all sectors and has new demands. Robots are serving food and sanitizing restaurants, and artificial intelligence (AI) can be seen everywhere around us. The impact of technology is reshaping jobs and formalizing new occupations in every business sector, and the hospitality sector has not been spared from this disruption. New job roles are appearing. Skill sets that used to be important are no longer the only requirements. Human resource managers are looking for skills on CV's that will matter in the future. The constant evolution in technology improves our living home to workstations beyond our expectations. Traditional skills will still play a role, but it will become necessary to know the technological world evolving rapidly around us. Advanced analytics, agile development, and data management seem to be the most critical skills required for future jobs. The backbone of AI in any company will be made up of skilled, intelligent minds capable of understanding business problems and applying AI to solve issues effectively. AI redefines the relationship between an employee and manager in an AI-driven workplace. A robotic might even become your line manager in future. The soft skills required to work successfully in an environment where teams are made up of robots and humans, working together will be different than required a few years ago. Over the last decade, technology has transformed many aspects of the hospitality industry. To be prepared for the futuristic requirements, it is mandatory to re-train and re-skill all those employees to make them skilled and fit the industry's altering needs. Chapter focuses on the jobs that will become hybrid and new job descriptions that will be created in the hospitality sector that will place more emphasis on developing human intelligence and critical thinking. It will also look at the educational institute's syllabi, and the new design will meet the future requirements.

6.1 INTRODUCTION

In the last couple of years, the world of work has changed forever. It is a brand-new day, and welcome to the new world of work in the hospitality sector. What used to be essential skills has become less attractive to HR managers in the past.

The job market will look very different in the next few years. No longer is work skills set in a tradition, but instead on the possibilities of what the future promise [1]. Future cities will have different requirements for working and preparing themselves for the future. Cities will be designed to meet one's basic needs in less than 15 minutes [2]. Cities will become smaller as much work can be done remotely, and many people prefer to live outside of them lately. Today's working environment is all about rising tech trends, and the pandemic was only an accelerator [3]. Consumer behavior and spending are changing. The hospitality sector is adapting to new rules and values, which will affect our relationships with customers. Work locations have changed; either employee work from home if not required in the office daily, or the office layout has changed to meet the new standards in health and safety. Employee's dress differently. There is no need to buy expensive office attire if the only part visible is the face and shoulders of a person. Time management has been taken to a whole new level [4]. Family and work life are all intertwined. Skill sets have become rare or obsolete, and new skills are replacing them. In every sector today, change is inevitable, and humans need to be willing to accept change and skill ourselves in different ways. Employees are also more educated in various skills due to hours of online training for the past year.

The hospitality sector has changed somewhat differently from the other sectors. Many chefs are not returning to their old jobs in the food production facilities, and the shortage of skilled staff is increasing. According to the Washington Post, an Amazon warehouse hourly paid employee may earn $17 per hour compared to a line cook who usually earns only $14 per hour, where he/she need to learn a skill to cook in the stressful, intense, and hot environment then why anyone would be doing cook's job [5].

6.2 FRONT OF HOUSE OPERATIONS

Pepper is the latest addition to the hotel and works as a concierge, answering guests' questions and entertaining guests. In a research project, "Pepper robot movements, gestures, and conversation skills were controlled via a real live agent. The idea of the study was to establish what guests think about the quality of service. Hotel guests gladly accepted the pepper robot, and most guests would use Pepper again in the future" [6]. The implementation of technology in its early years worried many hoteliers, specifically about the online check-in and checkout, that it may spoil the guest's experience. However, it went in vain since it raised guests' service standards and comfort [7]. Today guest check-in procedure is not limited to the desk, mobile device goes along with the guest anywhere in the lobby. Guests can also book into a hotel from whatever device they have in hand. The sluggish procedure and heterogenous service to each guests check-in has moved almost "90% of millennials to their mobile phone applications for smooth and faster personalized experience." [8].

Cloud computing empowers hotels to be ready to welcome guests and fulfill every individualistic demand from special dietary restrictions, for example, allergy requirements to personal choices when it comes to the preference of a particular room [9]. Imagine the guest arriving at the hotel; before he walks into the hotel, all the required data about his preferences and health issues are already recorded on a cloud system. In this way, the guest can be allocated a specific room, and amenities can be placed in the room according to the customer's special needs. Social distance can be kept between the hotel and the guest.

When the guest arrives later that day in the hotel restaurant, a menu card will contain only dishes suitable for the guest to eat. The electronic menu has been customized to update in a few seconds to meet the diet restrictions of the guest. Front of house will have more free time to assist guests with more important questions and add a warm human touch when the guest arrives at the hotel. The hotel will be upskilled to deal with demanding guests most effectively. All hotel employees will be trained in cultural diversity and how to deal with any situation empathetically. At many hotels worldwide, the guest will find the latest technology assisting the hotel in different ways, improving customer service, and meeting all stakeholders' requirements.

6.2.1 THE LATEST TECHNOLOGY IN FRONT OF HOUSE OPERATIONS

The past year has opened new opportunities for employees in the hospitality sector. The acceleration of technology implementation in this industry has enhanced possibilities for becoming more specialized in a specific area and learning new skills of the future. The following technological advancements have redefined hospitality industry [10]:

- Cloud computing;
- Service automation through AI;
- Integrated guest applications;
- Going touchless in all processes;
- Wireless mobile devices checking in guests;
- Location-based services;
- Hi-tech lounges;
- Self-service meeting spaces.

Today, guests understand that implementing robotics in the hospitality sector is not to replace expensive robots to save on costs but rather to have secure and real-time execution for each guest. The experience of a guest begins from brand awareness, and it continues with the knowledge of the brand and escalates throughout their entire journey [11]. The guest will expect a seamless experience not filled with ambiguous erroneous data mismatch between integrated systems.

The diagram (Figure 6.1) demonstrates that minimal human contact is happening. Not only is it meeting the health and safety requirements of today, but when a hotel is applying automated check-in procedures, the overall check-in time will be reduced. No longer will guests have to wait in long queues, especially during peak times. Hotel staff will have more time to pay attention to customers' immediate needs. Employees will learn additional skills to enhance the overall customer experience, such as handling demanding customers or understanding guests with a different cultural background. Some jobs will be eliminated and replaced by robots as it is more cost-effective over time. Some jobs will never be fully operational without the assistance of a human being, and there are also specific skills that only a human staff member can execute successfully. In Figure 6.1, the diagram demonstrates the outlay of a hotel lobby today. Already, many hotels have implemented some or most of this technology.

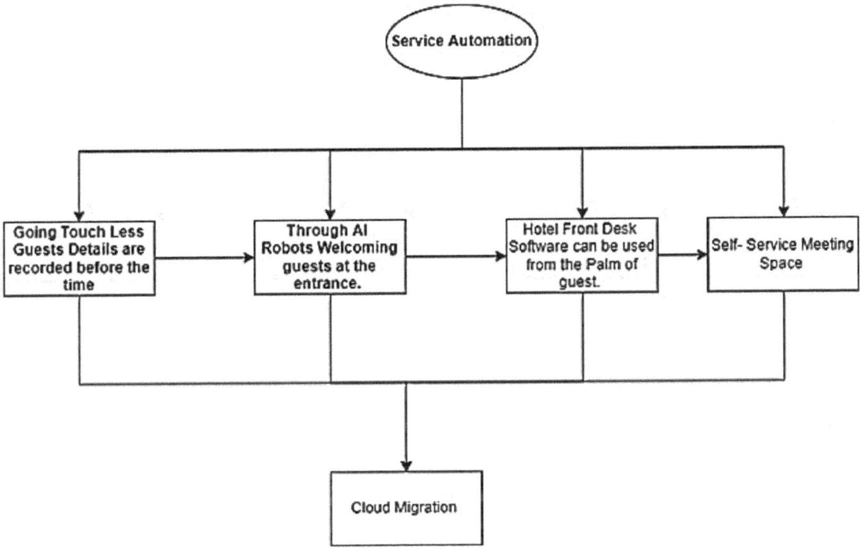

FIGURE 6.1 Design of a hotels' lobby with the latest technology (diagram created by the researcher).

Most hoteliers come from a non-technical background and, therefore, see robots as a threat [12]. Robotization will not only be taking jobs but also create new occupations. Many new occupations will arise, occupations we had not even thought of this very moment. Any job that includes critical, strategic, or creative thinking will not be replaced in totality by robots. Human interaction will be required in these types of jobs, and robotics will rather be part of a team. In Table 6.1, a comparison is made between replacement jobs and robots and automation in the hospitality sector. Staff members will require future skills to be successful in their job.

TABLE 6.1 A Comparison of Jobs and Robots in the Future

Old Job Description	Robot/Automation Replacement	Human Skills That is Irreplaceable or is Required in This Job Description	Replacement Jobs for Humans
Baker	Bread bot [13]	Creativity	Recipe developer
–	Mechanized mixologist [14]	Emotional connection/ creativity/empathy	Cleaners/stock replacement
Barista	Barista robot [15]	Emotional connection/ creativity	Consultant barista that teaches robots
Busser	Service robot [16]	None; the less human contact, the better	Cleaner
Butler	Bot Handy or Botlr a robot butler [17]	Empathy	Human butlers are more cost effective.
Catering manager	Advanced CRM system	Complex functionality of the brain/Emotional connection.	Robots and automation will assist managers in their jobs.
Commis chef	Collaborative robot [18]	Perfecting flavors and presentations.	Commis chefs will learn from robots.
Dishwasher	Dishcraft robotic [19]	None, as this robot is cleaning, washing, and sterilizing the crockery and cutlery.	Assistant will be required to sort the dishes and push the trolley to the dishwasher.
Event planner	Automation and robots to assist with different tasks.	Human touch and experience cannot be replaced by a robot.	Robots and automation will assist in creating a leaner operation.
Executive/ sous chef	Bot chef [17] and electronic tongue [20]	Creativity/complex functionality of the brain/ emotional connection.	A robot can assist with prep work, and can taste food.
Food and beverage manager	Advanced CRM system	Complex functionality of the brain/Emotional connection.	Robots and automation will assist managers in their jobs.
Food expeditor	Teachable robots [21]	Finesse and creativity	Robots can perform quality checks on its own.
Food runner	GogoBot, a food delivery service bot [22].	Critical thinking and creativity.	Robots can set up tables and the human must design the outlay of the table.
Guest relation officer	Pepper robot [23]	Emotional intelligence, effective communication, human emotions.	Humans make eye contact with a customer; therefore, humans should be mixed with robotics.
Guest relations manager	Advanced CRM systems	Complex functionality of the brain/Emotional connection.	Robots will be an extension for certain tasks.

TABLE 6.1 *(Continued)*

Old Job Description	Robot/Automation Replacement	Human Skills That is Irreplaceable or is Required in This Job Description	Replacement Jobs for Humans
Hotel concierge	Pepper robot [23]	Human emotions, empathy	Robots will be an extension of certain tasks.
Hotel front desk agent	Sanbot [24]	Empathy and emotional intelligence	Robots will make humans obsolete in this function.
Hotel HR manager	Advanced CRM systems	Complex functionality of the brain/Emotional connection.	Robots will be an extension of certain tasks.
Hotel maintenance manager job	Spot [25]	Complex functionality of the brain/emotional connection.	Robots will assist with maintenance.
Hotel manager	Advanced CRM systems	Complex functionality of the brain/emotional connection.	Robots will be an extension for certain tasks.
Housekeeper	Cloud robots and smart technology [26]	Critical thinking skills	Robots will replace humans for hygiene.
Pastry chef	LR mate [27]	Creativity	The robot can work without a human being. The robot has a reach and dexterity of a human arm.
Porter job	Robo-porter [28]	Empathy	Robots will not require any human assistance in this type of job.
Recipe developer	AI algorithms	Creativity	Robots can assist in tasting the food.
Room Attendant	Rosie the robot [29] or maid bot [30]	Robots can clean far better than their human counterparts.	Human staff focus on guest satisfaction tasks.
Waiter/ waitress	GogoBot, a food delivery service robot [22]	Empathy, emotional intelligence	Robots will assist waiters.

Source: Created by the Researcher.

Robots and automation will assist human beings in their daily tasks. Through further research, it is already proven that humans and robots working together in a working environment improves efficiency [31]. Hospitality staff will require different training in future, such as:

1. **Advanced Hospitality Skills:** The hospitality skills are real talents required to accomplish in the process of hospitality services and can be split up into hard and soft skills:

2. **Hard Hospitality Skills:** The skill is limited to tasks or essentials needed to complete technical tasks. It can include culinary skills in back-of-house operations or skills required to operate a reservation system [32].
3. **Soft Hospitality Skills:** Soft skills are social behavior requirements needed to work or manage employees in the hospitality sector, such as cultural understanding and emotional intelligence [32].

In the future, hard skills will not be long enough to succeed in the hospitality sector. Employers are looking for hospitality with both hard and soft skills [33]. It was acceptable for kitchen leadership to be highly talented in preparing food only in the past. Many staff accepted the lack of empathy, disrespect against cultural issues, lack of communication, and leadership [34]. It resulted in fewer young people seeing themselves in a career that did not respect them. Today, developing cultural awareness, multitasking, and communications skills is paramount to delivering outstanding customer experiences. The increasing complexity of this sector is leading to brand new job profiles and is forcing hospitality schools to bring in associated classroom training for future staffs [35].

A chef should cook delicious food and be skilled in asset management, sustainability, and social responsibility. Chefs have a vital role to play when it comes to the rebalancing protein consumption of the future. Future chefs must be equipped with skills to create tasty, nutritious, and sustainable food. Knowledge about environmental issues will be part of learning [36]. Teaching the correct curriculum in culinary schools will prepare young chefs for a successful career that will work with the latest technology and apply future skills effectively. Empathy and emotional intelligence will play an important role when dealing with different cultural groups. Emotionally intelligent ones can easily identify their emotions. The hospitality sector is well-known for cultural diversity, and it is an environment where you deal with all types of personalities [37]. Therefore, emotional intelligence is a valuable skill for understanding and managing guest expectations and a team without affecting the normal flow. Not everyone is emotionally intelligent. However, this skill can hone into a person through continuous training. Staff with higher emotional intelligence makes better leaders. Research studies have also proven that female leaders have higher emotional intelligence than men. Leadership belongs to women in future [38]. Food production facilities are generally male-dominated and often lack emotional intelligence. Many staff members suffer from anxiety, stress, and depression. If management is trained to deal effectively with mental health issues, it can improve staff's overall well-being and contribute to a happier workplace. Creativity and innovation are considered key competencies in the future of work and are not fully replaceable by robotics. Ai-Da is the latest addition to robots that can create art.

The Ai-Da an almost realistic robot getting better day by day [39]. Sugar artists might have strong competition in the future, and maybe you will require exceptional creativity to keep up with robots in the creative area. Other skills that will be very important to have on your CV will be flexibility, networking, organization, commitment, and a can-do attitude. Emerging technologies such as AI and robotics will have a strong influence on the future skill set. Skills that will become important are cloud computing, and the demand for cloud skills is increasing the need for specialists [9]. Another valuable soft skill will be analytical reasoning. Someone will be required to analyze all the big data that we require. A qualification in Lean Six Sigma will be helpful to have to create lean cultures in the hospitality sector [40]. People management who can coach and empower staff will be a valuable skill. Micromanagement will be a less desirable trademark as the command is overtaken by the coaching and control is distributed with empowerment [41]. It will also be a valuable skill if your manager knows how to deal with mental well-being and recognize depression, stress, and anxiety before it becomes an issue in the working environment [42]. Not only will collaborative robots play an essential role in the hospitality sector, but collaboration between humans or humans to robots will help make projects or work results more successful [31].

We learned in the past that how quickly life can change; being adaptable to changing situations and expectations that might not have been met make you more valuable to employees. Especially in the hospitality market, this will be a skill that will require practice and becomes a part of the employee's DNA. Without time management, a hotel operation will not be able to give 100% customers satisfaction or have a high-efficiency rate. The guest of today wants everything promptly. A well-organized team can achieve this. Managing time requires discipline, planning, and effective ways to manage stress levels. Not only is this a valuable skill for the hospitality industry, but it can be applied in all other parts of your life.

Oral and written language skills will become a rare skill in future as lately everyone is working on technology and people forget how to write. Due to social media "slang" language, the correct grammar and spelling are also creating challenges. The ability to listen to what the person opposite you is trying to tell you is scary. Very few people have the ability to listen and to understand. Most people try to think of a suitable answer before the other person's communication. The ability to listen to understand needs to be practiced [43].

6.3 MANAGEMENT SKILLS OF THE FUTURE

Developing managerial skills for future management roles in the hospitality sector will be for all professionals. What skills will play an important role in future in management?

- Statistics evaluation and measurement skills;
- Critical decision-making skills;
- Cultural diversity management;
- Structural design skills;
- Planned progress skills;
- Hospitality marketing skills;
- Fiscal analysis and economical management skills;
- Property management skills;
- IT skills;
- Effective presentation skills;
- Cognitive flexibility;
- Quantum computing skills.

The rise of digital technologies and robotics means that the employee will have to handle many adjustments. Leaders will have to quickly prepare their mindset for the multiple complex ideas all at once and deploy strategies for operational excellence. Showing qualities of advanced multi-tasking will be highly valued by employers. Being digitally literate will be another valuable skill to be added to your CV. Finding, evaluating, and communicating information correctly through various platforms will enhance opportunities. This will include grammar, composition, and producing text and creating images using the latest technology. The 21st century generation must be trained to handle the upcoming novel challenges. Humanity should be prioritized before profits and power [44]. A diploma in hospitality management or culinary skills will eventually no longer be enough to build a successful career in the hospitality sector. A master's degree in AI can help advance your digital skill set and accelerate your technical skills in these emerging fields. As we have already seen, not all skills will be replaced by robots or the latest technology. Robots might assist in repetitive, boring, and dangerous work or even office work, but humans will still handle judgment and decision-making. Robots and automation technology will be calculations; however, humans will do data analyses to get the most effective end results. According to World Economic Forum, robots and automation will create more jobs in future, but with different job descriptions [45]. It would help if you learned to think outside the box. Hospitality employees need to think of creative solutions when issues arise.

6.4 TECHNOLOGY WILL CHANGE THE WORKING ENVIRONMENT

6.4.1 SPEECH SEARCH AND CONTROL (INTELLIGENT VOICE INTERFACES)

Voice search technology is the latest trend in the hospitality sector, and it will most likely stay with this industry. Customers today are using voice search on their mobile phones to find accommodation, restaurants, and attractions [46]. The guest will ask a question by speaking to a voice assistant on the phone and get almost immediate results. The demand for voice control is growing around the world. Voice control can be used within the hotel rooms, and services can be availed throughout with equipped devices, such as closing the curtains or dimming the lights. In restaurants, automated ordering means that customers may place their orders instantly. Guests can place their order via a speaker, and the order will be forwarded to the kitchen. Order takers will be eliminated from future jobs. Instead, they will assist customers where difficult choices must be made between the different menu items available. Not all food companies were ready for this type of service. For many companies in the hospitality sector, COVID-19 has pushed the industry to implement app-based service seeker's paradise. Implementing technologies that let guests operate the room temperature, place an order, or book an excursion. Room services will be essential factors that will help customers to choose from and distinguish the right choice from their competitors. Hotels should be capitalizing on this. Websites will have to be set up in such a way to make sure that voice search can be applied and used correctly.

6.4.2 MARKETING DIGITALIZATION

The branding and promotion have been consistent at each touchpoint, or customer interactions could lead to more and more booking and revenues. The waiters often see a restaurant promotion for the first time only when guests ask for this. Quality control and consistency of marketing material will become very important. All the social media platforms will have to advertise the same product across all platforms. The requirements for the marketing job descriptions will change. The new job descriptions will move from creative to data-orientated. The marketing team will include mood and empathy manager, six sense analyst, purpose planner and data ethnographers [47]. The Mood and Empathy manager will analyze how guests feel about the stay at your hotel or the visit to your restaurant. Surveys on digital platforms about the guest's experience when staying at your hotel will become outdated. A neuro A/B tester [48] will record the brain activity of the guest when staying at your hotel [47]. The job of marketers will become far more complex, as guests are very well informed and know exactly what they want to get. A successful marketer will have an agile mindset, including adapting to consumer dynamics that can immediately change if the need arises.

6.4.3 BACK OF HOUSE TECHNOLOGY

More robots are being utilized in food production facilities around the globe. In two years, the implementation of collaborative robots will almost double in the food industry. Figure 6.2 illustrates the expected implementation rate of collaborative robots in the food industry [49].

Up to this moment, most culinary schools were offering training to aspiring chefs on the hard skills required to complete a task in the kitchen well. The concern is when robots move into the kitchen. Boring, repetitive tasks will be replaced by robots or some form of technology. If culinary students are not prepared with skills that will matter in the future, they will fall behind and not contribute successfully to a food production facility.

A sixth sense analyst will be required in the marketing department and be beneficial to convert AI-driven predictive intelligence into customer's feedback insights. Basic quality control will no longer be completed by a human counterpart but rather by a robot that can measure the food temperature, determine any traces of food bacteria and monitor standard recommended plating when food leaves the kitchen. COVID-19 has accelerated many of these jobs into a new description or completely removed it. People might be not so eager to return to pubs and bars due to the fear of enclosed spaces. "While there seems to be a new video every day off with youth blithely partying outside (and inside) bars without masks, many people have been drinking less during the pandemic" [52]. The concept of nightclubs, where many people are sometimes cramped into small spaces drinking and dancing in an unhealthy atmosphere, is changing. Teammates are needed to maintain stock levels in automated bars. No human being is no longer required when a robot collects dirty dishes in a restaurant. A Japanese company has developed a robot called ACUR-C, which can serve and bring back all F&Bs without any human assistance [53].

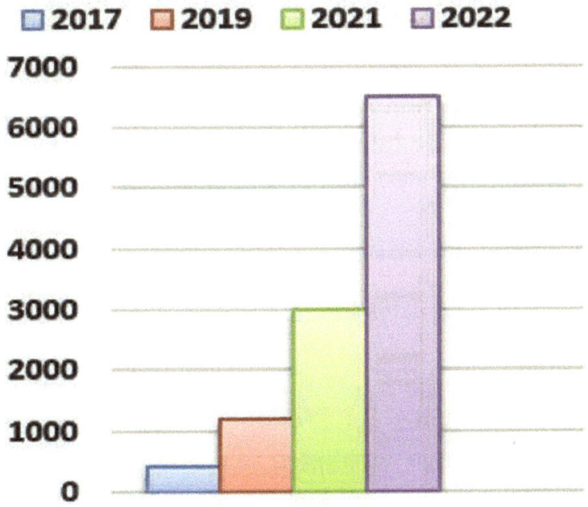

FIGURE 6.2 Collaborative robots in the food industry [51].

Event planners will never be entirely replaced by robots, as a robot cannot give that human touch or years of experience, which is learned through a lifetime of trial and error. Robots will streamline and assist with tasks, making the event planner's job much more manageable. However, an eye for detail and the final touch can only be completed by a human. Robots can taste different foods today. Chinese food manufacturers use.

Robots to ensure the taste of the recipes to be intact throughout in the bulk production [50]. The robots, with their vision and nasal capabilities, are positioned at the production bays to assess the food quality. A robotic tongue would be an asset to any food production facility. The advantages will include doing quality checks on special dietary foods to ensure that they do not contain any substance that the customer might be allergic to, for example, gluten or lactose-free food. A robotic tongue would prevent food production facilities from sneaking in harmful ingredients without mentioning them on the label [51]. A robot can be programmed to pick up ingredients that might harm humans. Robots can also be programmed to pick up when pathogens and toxins are present that could harm human beings when they consume food.

Teachable robots equipped with grippers and instrumentation performs quality checks in different ways. The ready to serve food will pass on a conveyer band and pass the robot. The robot will scan for temperature and ensure that the food is leaving the kitchen at the correct temperature. The robot can learn different plating ideas. Once the food passes the robot, the camera will scan for uniformity and consistency. As technology improves daily, the only limit is the human imagination. There are many ways to check quality with automated equipment. Even a drone can be applied for quality checks [52].

6.4.4 QUANTUM COMPUTING

Quantum computing is the most world-changing technology up to date, and it will integrate in future in our working environment in only a matter of time [53]. The quantum revolution will bring new hardware and new software. It will become the new standard to protect valuable information. According to experts in this field, it might take another 4–5 years before it will benefit a wider audience. The basic unit will move from bits to qubits [54]. Quantum computing will solve finance, materials, security, and transportation impossible computational tasks. Quantum computing will also find new ways to fight climate change. There is no practical way to remove gigatons of CO_2 (Carbon Dioxide) gas from the atmosphere. Quantum Computing Engineers will become a critical position for many companies in the future.

6.5 THE FUTURE OF WORK IN THE HOSPITALITY SECTOR

The job market is constantly changing, so changing jobs every three to four years is acceptable in today's market. Everyone in the hospitality sector has been affected by the COVID-19 crisis and has underlined the importance of travel and tourism interconnection with other industries. Not only will educators in the hospitality industry have a broad knowledge about the subject, but they have responsibility to pass it on to the new generation. The student shall learn how to be streetwise by not only learning a language as a subject. But also learn how to solve an issue or prepare mentally for the job. Future hospitality will need a broad knowledge base to succeed in their job. Robots and automation will assist in education, but a personal touch will still be required from time to time. Specialist skills will be very important. Your guests will search for authenticity, the real product, coffee beans grounded on the premises, and bread baked fresh from the oven and served immediately. Health spas will always require a human to add the personal touch, to show empathy and hospitality to guests. Eventually, technology and robots will impact all manual labor. Any job that needs artistic flair and creativity will only be assisted by robots, as robots cannot replace creativity. Robots will not fully replace cake artists and artisan bread makers.

The room attendant that can fold a towel in a creative decoration to welcome guests will also not be possible to replace by a robot. Chefs and bartenders will always be in demand because of their creativity, flexibility, and complex management skills. Salespeople cannot successfully replace robots, as all salespeople have their unique styles. A salesperson can be trained on the basic skills; however, a deal is close in how the salesperson reads the customer and approaches the customer. Flexibility and management skills recruiters will be the sought-after skills HR managers will be looking for. Employers value skills over experience, and by keen observation and attainment of highly desired skills, seekers may always remain in demand among employers.

6.6 JOBS OF THE FUTURE

To start a successful career in the hospitality industry, future skills and practical experiences to analyze evolutions and tackle the challenges attached to industry trends will have to be developed. In some jobs, certificates, and accreditation will become less critical. Instead, it will be about the skillset that potential employees can offer than a file full of certificates. Jobs that will be in demand in future will be CRO (chief robotic officers), robotic engineers, robotic programmers and specialists involved in building and maintaining robots. Robot specialists will be required in the hospitality industry to look after robots and collaborative robots that will form part of a team. It will also be essential to have a Solar Photovoltaic Installer in the maintenance team [55]. This person maintains and service solar panels and solar panel systems and troubleshoot when there is an issue. Individuals need a high school diploma to work in this profession, but most of the skills needed are learned through on-the-job training. AI is changing how companies work, and robots supervise their human teammates. Companies are replacing human labor with machines. Fanuc in China is such a company [56]. It is a fully automated company that does not need human supervision at any stage of its operation. Intelligent systems are replacing humans in the manufacturing sector and service delivery. Moreover, due to this, human workers are moved towards lower-paid jobs or makes them unemployed. Our workforce will change entirely in the future. The hospitality sector will adopt Lean Six Sigma Skills from other industries, becoming a crucial part. New job descriptions that can arrive from this could be Chief Robotics Officer [57], Lean Specialist, Continuous Improvement Manager, and a Quantum Computing Engineer [58]. The food safety and hygiene specialist role will change and become more of an advisory role.

6.7 HOSPITALITY JOBS THAT WILL BE ELIMINATED IN FUTURE

Virtual reality (VR) will enhance travel agent jobs; the destination can be explored in detail. See what the hotel accommodation will look like and what activities await the tourist. Robots will replace fast-food restaurants. Most fast-food recipes are repetitive work that a robot can easily manage. KFC has already opened its first shop, mainly by robots [59].

6.7.1 THE EMPLOYEE OF THE FUTURE

In the past, a job was security, a place where you could accelerate in your career or download your artistic skills, a place where you could share your work with your colleagues and share burdens. All of this has changed due to the pandemic. According to Maslow's pyramid, right at the bottom is safety, and depending on where you come from, it was not something the average person paid much attention to [60]. We have seen nature return to cities during the past few months, and thc air has become cleaner. Offices will have to adhere to safety protocols. Safety is not only related to cleanliness, but employees must also feel safe in terms of psychological wellbeing. Mental health has suffered through the pandemic. Most managers were not trained in handling employees exposed to prolonged challenges and chronic stress, which can change the brain's architecture, causing fatigue, anxiety, fear, and depression [61]. With these conditions, a feeling of emotional safety and wellbeing is critical to people's ability to participate fully in their work. Leaders need to be trained on how to provide social safety in which employees feel levels of trust and inclusivity [42]. Sadly, many leaders today lack these new skills and would instead terminate employees than assist them in emotional security, employment welfare, and social stability. In the new working world, managers need to manage the whole person and not only work. Before the pandemic, many employees were at a different place in the Maslow Pyramid. More than basic needs were met.

Love and being loved to play an essential part in psychological needs. Today, many employees did not see their families due to this pandemic for a very long time. Some of them have lost loved ones. Working from home has limited our time with our immediate family. People work the whole day primarily, with no specific times in place [62]. Even if your love needs have been met, human beings need to socialize. Groups of family and friends have changed, and most people today struggle with anxiety, depression, and stress. Some days, you do not feel like talking to people due to anxiety. Esteem needs and accomplishing something are the fourth level in Maslow's hierarchy, including self-worth and respect. If the manager in the company is not skilled to lead but rather to manage, the self-worth of many employees will be destroyed.

Mental health plays a vital role in thc life of employees today. Not only are there the different needs mentioned by Maslow, but what about the constant fear created through the latest technology, overloaded workforces, and above-average expectations of employees? For many companies, productivity went up due to people working longer hours from home, raising productivity and, at the same time, anxiety. Therefore, unless leaders can address the causes of worker anxiety and recognize when employees are reaching a burnout stage, we are heading to a very unsustainable future [63]. Anxiety reduces job satisfaction, negatively affects colleagues' interpersonal relationships and decreases work performance [64]. Flexibility will be essential for employees. No one wants to spend most of their time at work. Employees demand flexibility with the significant disruption that the pandemic has caused in their private lives. Changed school schedules, many students today are getting online training, limited daycare options, and working different programs are just a few examples that can weigh heavy on employees and require a lot of their attention and time. People also value to own time and has learned that to live in abundance requires minimal effort.

6.8 THE NEW TRAVEL AGENT

Before the pandemic, customers used to make their travel arrangements. It is easy to search for flights, find hotel accommodation that is the most suitable on a booking platform, and arrange to transport from the airport to a hotel through an email. The role once again of the travel agent has changed after the pandemic. Today, customers want one reliable person who has a broad knowledge about the destination they are about to visit to assist with everything. The new job is a travel consultant that can arrange form the quarantine period, biometrics, required documentation to enter a specific country, suitable hotel accommodation and the transportation. Customers are looking for an all-in-one package. The travel consultant must have previous destination experience to provide the best options possible to the customer if the travel consultant can arrange for biometrics and fingerprints to be taken at the customer's location, the better. To provide all the new requirements, students will have to be trained in different subjects in Hotel and Tourism Management Hotel Schools.

6.9 THE TOUR GUIDE OF THE FUTURE

Many citizens in different countries make a living out of being tour guides. There was never a specific criterion for what a tour guide should do in many countries or what qualifications he should have. Especially in former USSR countries or third-world countries, the only requirement was to have a broad knowledge of your own country. Tour guides often try to rip you off when you visit some of these countries, especially if they can identify that you are from another country. Often, many tourists were cramped into one boat or one bus to make additional money. This will all change in future. More people will prefer private tours, only one family or a few friends. It seems that tour guides will charge much more for a single family or group of friends in the future. The focus will also be on low-density tourist attractions. This means that tour guides will have to determine when will be the best time to visit the site; once again, the working hours of the tour guide will change to be able to meet these requirements. More people will prefer to travel locally. Tour guides will develop innovative ideas to showcase their product to local people. Out of the box skills, creative, and critical thinking skills will be necessary. Value for money offered by tour guides will increase their business. Portraying something unique in a well-known area will increase money-making opportunities for tour guides. No longer can tour guides be set in a routine. As all of us have learned over the past few months, life can change overnight; Tour guides will have to be flexible on pre-bookings. If a customer cannot attend a certain pre-booked tour, options to move the tour to another date should be offered. Tour guides will also have to figure out how to keep everyone safe and make sure visitors follow health and safety protocols in every situation. Before the pandemic, tour guides struggled to maintain an income due to the many self-guided tours everywhere. Tourists or customers will prefer to see a new country, with someone familiar with all the health and safety restrictions, and then travel independently.

6.10 HOSPITALITY MANAGEMENT CURRICULUMS OF THE FUTURE

Educational institutes will have to adjust curriculums to include robotics majors and chief robotic officer skills. Robotics offers numerous opportunities, ranging from engineering to psychology, computer science and kinesiology. Theoretical knowledge will be essential to future careers, and different projects will be completed throughout the year. Students will have to develop entrepreneurial business projects and create a virtual company. Internships will improve the skill set of the student.

6.10.1 THE NEW CURRICULUM FOR HOSPITALITY MANAGEMENT

Hotel Management Curriculums will have to change to adapt to the future of work in hotel management, as depicted in Table 6.2.

TABLE 6.2 Future Work and Discussion on Hotel Management

Hotel Management Level 1	Hotel Management Level 2
Tourism and virtual reality	Space tourism
Cultural studies	Cultural studies in business management
Room service 1	Room service 2
Food and beverage service 1	Food and beverage service 2
Mandarin Chinese basic	Mandarin Chinese for business
Soft skills of the future 1	Soft skills of the future 2
Basics of lean culture	Continuous improvement
Sustainable hotel management	Strategic leadership management
Data Analytics	Robotics and Automation Level 1

In the future, potential customers would like to experience a simulation of a location before they take out their hard-earned cash to book a vacation. VR or virtual reality refers to interactive images with a clear 360° view of a location [65]. Instead of teaching future hotel managers only about tourism, the student should know about VR tourism and how to use this in the most effective way to promote a destination that allows the customer the experience of being there [65]. Future hotel managers should learn about VR tourism, but they should also learn about Space tourism, a new traveling area that will form a part of customers' requirements [66]. Management will have to be very knowledgeable about the new places to explore.

All hotel management courses should include cultural studies and management education [67]. It will no longer be important to learn only about the country or environment where the educational facilities are located but should teach students about the different cultures around the world. It will enhance the student's cultural knowledge and be beneficial to manage a culturally diverse team. To be an exceptional leader, a manager must motivate a team to bring the best out of them. Currently, Mandarin is the most spoken language around the world. It is estimated that 917 million people speak Mandarin worldwide [68]. With the continuous rise of China as a global power, Chinese, like English, will become a global language. Students will learn Chinese, which will put them one step ahead of their counterparts in the hospitality sector. French, Spanish, and English could be offered as an additional subject [68]. It could also be offered to students who cannot speak English. Currently, most classes in International Hotel Management Schools are offered in English, which will improve the English vocabulary. The critical skill for the decision-making positions is Effective decision making [69]. Some people are not suitable for leadership roles due to their inability to make effective decisions or any decision at all. Not only must the person be able to make effective decisions, but they must be capable of implementing them. Decision-making skills can be divided into critical problem solving, collaboration not only with humans but also with robotics, emotional intelligence and sound logical reasoning.

Lean Six Sigma principles can play a pivotal role in the Hotel Management and Tourism sector [70]. Lean Six Sigma combines two concepts, lean and Six Sigma, into a single management framework that focuses on eliminating waste, reducing defects, and improving overall efficiency [71]. These principles should be looked at as robotics and technology will be an essential part of this industry in the future and could play an important part in the tourism industry. It could be applied in food and beverage (F&B) services, gambling, and the recreational sector. In the accommodation sector, it can help to maintain room inventory and create an effective supply chain. It will be necessary for a student to learn and understand the benefits of Lean Six Sigma and how this is linked to customer satisfaction and leads to customer or guest loyalty.

Sustainability is becoming a significant concern for public and private sectors worldwide, and the hotel management and tourism sector is no exception [36]. During this environmental course, the student will develop important skills to understand our planet, the limitations humans are creating, and their important role in sustainability, ethics, and sustainable ecotourism management [72]. As exceptional leaders of the future, the students will learn how to align business operations with issues around climate change and creating greenhouse gas emissions. Ethical issues around the working environment should also be addressed, as many employees suffer after the pandemic. Managers must be taught mental health and how to deal with employees during difficult stages [73]. The customer that is traveling today or soon is seeking sustainable travel. They are also looking for companies with a strong environmental reputation and policy. Strategic Leadership Management will prepare students to envision a strategic vision for an organization. They will understand the importance of stakeholders and how to bring them together. Flexibility skills and adapting very quickly will form a part of the course. They will also learn to understand the complex relationship between an organization and its environment. Effective decision making can only happen through efficient data analysis and is, therefore, necessary in any organization to determine its position in the market. Data analysis will help to understand the issues that the organization is facing. Data on its own is only numbers and figures. Once the manager understands that the data process can be streamlined, money can be saved, and net profit on financial statements can improve. Data is another word for hardcore facts, and if the manager can read and analyze the data effectively, it will give the actual state of the organization [74]. Robotics and automation have become an essential part of the hospitality industry. Research has proven that service robots and collaborative robots will increase efficiency and improve productivity [31]. When a robot forms a part of the team, managers will have to show how robots work and what is required to keep them working. To save any automation or robotics going, they will require regular maintenance. Hotel management will become tech driven. It will be important that the leader or the manager understands technology.

6.11 THE NEW CURRICULUM FOR CULINARY ARTS

Culinary arts students in traditional schools learn mostly about hardcore skills. To be effective in back-end operations, curriculums should be changed to improve the future of chefs. Labor issues in food production facilities were an issue even before the pandemic. Many chefs are suffering from anxiety, depression, and stress. Leadership in these facilities is not trained to deal with mental health issues or find solutions for the ongoing issues. The following subjects are recommended for a culinary arts program, as stated in Table 6.3.

Currently, after the pandemic, the labor issues have exceled. During the pandemic, many chefs have reskilled themselves in other areas. The labor shortage has increased, as well as mental issues. Robots, or rather collaborative robots, will become more important in this industry

TABLE 6.3 Culinary Arts Program

Culinary Arts Level 1	Culinary Arts Level 2
Food hygiene and food safety	Automation and food safety
Basic cooking techniques	Advanced cooking techniques
Soft skills of the future 1	Soft skills of the future 2
Basics of Lean Six Sigma 1	Continuous improvement
Sustainability management	Strategic change management
Data Analytics	Robotics and Automation Level 1
Creative arts and design	Entrepreneurship and business studies

to replace the lack of human labor [18]. Already, robots form a part of the team. Does the question remain if the implementation of robots will improve the overall wellbeing of a chef? Some research in other sectors already suggests that humans prefer robots as their teammates [75]. Future will unveil the reality.

6.12 CONCLUSION

The pandemic has changed our world forever. Our offices will become more technology-driven, from how we dress to the hours we work. The change has either highlighted underdeveloped areas or changed how work is done. Robots will not take our jobs but rather assist us with repetitive tasks or daily tasks. Effective HRI (human robotic interaction) will play an essential role in future. Humans will become more creative and will be able to make informed decisions. Staff in the hospitality industry will give detailed attention to guests, enhancing guests' experiences and increasing customer satisfaction. Educational institutes will integrate skills that will improve the life of hospitality staff in many ways. Hospitality staff will be better cared for, and the well-being of staff will become necessary. No longer will the team be in a position where work will become repetitive and not stimulate the mind. Street smart, to know something about everything will become the norm. Staff will become cleverer. Skillsets will no longer be suitable for one sector but will be transferable to many other industries.

KEYWORDS

- **artificial intelligence**
- **culinary arts**
- **digitalization**
- **educational programs**
- **future employment**
- **hospitality**
- **intelligent voice interfaces**
- **job skills**

REFERENCES

1. Pate, D. P., (2020). *The Skills Companies Need Most in 2020—And How to Learn Them*. [Online]. Available: https://www.linkedin.com/pulse/skills-companies-need-most-2020and-how-learn-them-lazzaroni-pate (accessed on 07 September 2023).
2. Jenniferliu, L., (2020). *These 10 Small Cities Could Become the Next Work-From-Home Destinations*. [Online]. Available: https://www.cnbc.com (accessed on 07 September 2023).
3. Barati, L., (2020). *How Technology Will Reshape the Hotel Industry Post-COVID-19 Pandemic*. [Online]. Available: https://www.travelpulse.com (accessed on 07 September 2023).
4. Dierdorff, E. C., (2021). Time management is about more than life hacks. *Harvard Business Review*.
5. Riley, L., (2021). *As His Restaurants' Customers Return, a Miami Chef is Missing a Critical Ingredient: Workers*. The Washington Post.
6. Sajadi, R. E., (2020). *Meet Pepper: The Humanoid Robot Concierge*. EHL Insights, Lausanne Switzerland.
7. Ivanov, S., Seyitoğlu, F., & Markova, M. (2020). Hotel managers' perceptions towards the use of robots: a mixed-methods approach. *Information Technology & Tourism, 22*, 505–535.
8. P3 Hotels. (2021). *8 Reasons Why Hotels Must Offer Online Check-in* [2021 Updated]. [Online]. Available: https://www.p3hotels.com (accessed on 07 September 2023).
9. Hood, J. A. S. Y., (2021). *Cloud Computing in the Hospitality Industry: Explore How Hotel Analytics Supports Your Possible*. [Online]. Available: https://www2.deloitte.com (accessed on 07 September 2023).

10. Revfine, (2021). *The Latest Technology Trends in the Hospitality Industry*. [Online]. Available: https://www.revfine.com/technology-trends-hospitality-industry (accessed on 07 September 2023).
11. Manoukian, J., (2020). *How to Create a Seamless Guest Experience at Your Hotel*. [Online]. Available: https://crowdriff.com/ (accessed on 07 September 2023).
12. Starkov, M., (2021). *Are Robots Coming to a Hotel Near You*. Hospitality Net, p. Unknown.
13. Wilkinson Baking Company, (2020). The Mini Bakery: Shaking and baking things up. [Online]. Available: https://www.wilkinsonbaking.com/ (accessed on 07 September 2023).
14. Bradham, B. T., (2020). *Goodbye to Bartenders: Robots Could Soon Make Your Drink*. [Online]. Available: https://economictimes.indiatimes.com/ (accessed on 07 September 2023).
15. Rozum Café, (2020). *Five Best Robot Coffee Baristas*. [Online]. Available: https://www.roboticstomorrow.com/ (accessed on 07 September 2023).
16. Lu, V. N., Wirtz, J., Kunz, W. H., Paluch, S., Gruber, T., Martins, A., & Patterson, P. G. (2020). Service robots, customers and service employees: what can we learn from the academic literature and where are the gaps?. *Journal of Service Theory and Practice*, *30*(3), 361–391.
17. Samsung News Room, (2021). *Samsung Introduces Latest Innovations for a Better Normal at CES 2021*. [Online]. Available: https://news.samsung.com (accessed on 07 September 2023).
18. Universal Robots, (2016). *How Cobots Transform the Food Industry*. [Online]. Available: https://www.universal-robots.com/ (accessed on 07 September 2023).
19. Dischcraft, (2021). *Clean Dishes Every Time*. [Online]. Available: https://dishcraft.com/ (accessed on 07 September 2023).
20. Wu, L., (2019). *How Robot Tongues are Measuring Taste for the Food and Beverage Industry*. [Online]. Available: https://www.forbes.com (accessed on 07 September 2023).
21. Liu, Y., Li, Z., Liu, H., & Kan, Z., (2020). Skill transfer learning for autonomous robots and human–robot cooperation: A survey. *Artificial Intelligence, 128*(103515).
22. Mallick, M. M., (2021). *Gogobot*. Kiran Smart, Kuwait.
23. Softbank Robotics, (2021). *Pepper*. [Online]. Available: https://www.softbankrobotics.com/ (accessed on 07 September 2023).
24. Sanbot, (2021). Sanbot. [Online]. Available: http://en.sanbot.com/ (accessed on 07 September 2023).
25. Boston Dynamics, (2021). *Transformative Mobility*. [Online]. Available: https://www.bostondynamics.com/ (accessed on 07 September 2023).
26. Sørlie, I. E., (2021). *What Are the Benefits of Connecting Robots to the Cloud*. [Online]. Available: https://www.ericsson.com/ (accessed on 07 September 2023).
27. Fanuk, (2021). *Suteria Deploys Robots to Process and Decorate Cakes*. [Online]. Available: https://www.fanuc.eu (accessed on 07 September 2023).
28. Internet of Business, (2021). *Alibaba Launches Robo-Porter, Japan Hires Disabled People to Operate Robots*. [Online]. Available: https://internetofbusiness.com/ (accessed on 07 September 2023).

29. Technical University Munich, Rosie, (2021). [Online]. Available: https://robots.ieee.org/ (accessed on 07 September 2023).
30. Maidbot, (2021). *Think Outside the Bot.* [Online]. Available: https://maidbot.com/ (accessed on 07 September 2023).
31. Klumpp, M., Hesenius, M., Meyer, O., Ruiner, C., & Gruhn, V. (2019). Production logistics and human-computer interaction—state-of-the-art, challenges and requirements for the future. *The International Journal of Advanced Manufacturing Technology, 105*, 3691–3709.
32. Planday, (2019). *Hard and Soft Skills – What's the Difference.* [Online]. Available: https://www.instituteofhospitality.org (accessed on 07 September 2023).
33. Ellingrud, K., Gupta, R., & Salguero, J. (2020). Building the vital skills for the future of work in operations. *Eds. Sneader, K., Singhal, S., & Sternfels, B. What now*, 66–73.
34. Michael Page, (2018). *The Importance of Emotional Intelligence in the Workplace.* [Online]. Available: www.michaelpage.com (accessed on 07 September 2023).
35. Rizvi, A., (2019). *UAE Needs Vocational Courses – Not Just University Degrees, Say Parents and Teachers.* The National, p. Unknown.
36. Khatter, A., White, L., Pyke, J., & McGrath, M. (2021). Stakeholders' influence on environmental sustainability in the Australian hotel industry. *Sustainability, 13*(3), 1351.
37. Skonk, K., (2020). *How to Resolve Cultural Conflict: Overcoming Cultural Barriers at the Negotiation Table.* Daily Harvard.
38. Burchfield, S. K., (2016). Is the future female? How emotional intelligence and gender affect. *Human Dimensions of Organizations*, 1–48.
39. Brown, M., (2021). *Some People Feel Threatened: Face to Face with Ai-Da the Robot Artist.* The Guardian, p. Unknown.
40. Carvalho, C. S. C., (2016). *Integrating Ergonomics with Lean Six Sigma on a Meal Solutions Industrial Kitchen.* Universidade Nova de Lisboa, Lisbon.
41. Freedman, M., (2020). *How to Avoid Becoming a Micromanager.* Business News Daily.
42. DeTienne, K. B., Hooley, J., Larrocha, C., & Reay, A. (2020). How to manage an employee with depression. *Harvard business review.*
43. Eilers, C., (2021). *20+ Effective Communication Skills (Good for a Resume).* [Online]. Available: https://zety.com (accessed on 07 September 2023).
44. Stephanie, L., (2021). *Future Skills You'll Need in Your Career By 2030.* [Online]. Available: https://www.topuniversities.com/ (accessed on 07 September 2023).
45. Scribani, J., (2018). *Is a Robot Coming for Your Job.* [Online]. Available: https://www.weforum.org/ (accessed on 07 September 2023).
46. Lala, P., (2020). *Evolving Voice Recognition Technology Allows Hoteliers to Offer Guests Convenience Without Concern.* [Online]. Available: https://hospitalitytech.com (accessed on 07 September 2023).
47. Davis, E., (2019). *The New Marketing Jobs Which Will Emerge in the Next Decade – From Creative to Data.* [Online]. Available: https://marketingtechnews.net (accessed on 07 September 2023).
48. Klobas, R., (2021). *How A/B Testing Tools Work (A Guide for Marketers).* [Online]. Available: https://thegood.com/insights/a-b-testing-tools (accessed on 07 September 2023).

49. Universal Robots, (2021). *Collaborative Robot Market Forecast.* [Online]. Available: https://www.universal-robots.com (accessed on 07 September 2023).
50. Chen, S., (2019). *Artificial Intelligence Robots Boost Profits in Chinese Food Factories, Paving Way for Mass Use.* South China Morning Post.
51. Tarantola, A., (2019). *IBM Built a Robotic Tongue to Taste Test Hazardous Chemicals.* [Online]. Available: https://www.engadget.com/ (accessed on 07 September 2023).
52. Simonsen, R. B., (2021). *Drone Inspection in the Food Industry.* [Online]. Available: https://forcetechnology.com/ (accessed on 07 September 2023).
53. IBM, (2017). *What is Quantum Computing.* [Online]. Available: https://www.ibm.com/ (accessed on 07 September 2023).
54. CB Insights, (2021). *Quantum Computing Vs. Classical Computing in One Graphic.* [Online]. Available: https://www.cbinsights.com (accessed on 07 September 2023).
55. Revision Energy, (2021). *Solar Options for Inns, Restaurants, Hotels, B&Bs.* [Online]. Available: https://www.revisionenergy.com/ (accessed on 07 September 2023).
56. TrivediI, A., (2020). *The Threat of Robots Taking Our Jobs Gets Real.* The Japan Times.
57. Goncalez, C., (2019). *Who Manages the Robots? Introducing the Chief Robotics Officer.* [Online]. Available: https://www.machinedesign.com (accessed on 07 September 2023).
58. Ackerman, D., (2020). *Explained: Quantum Engineering.* MIT News, p. Unknown.
59. Drafting, (2020). *KFC Opens Restaurant without Employees, Based on 100% in Robotics.* [Online]. Available: https://www.diegocoquillat.com (accessed on 07 September 2023).
60. American Stroke Association, (2021). *Maslow Hierarchy of Needs.* [Online]. Available: https://www.stroke.org/ (accessed on 07 September 2023).
61. Ornek, O. K., & Esin, M. N. (2020). Effects of a work-related stress model based mental health promotion program on job stress, stress reactions and coping profiles of women workers: a control groups study. *BMC public health, 20*, 1–14.
62. Kost, D., (2020). *You're Right! You Are Working Longer and Attending More Meetings.* Harvard Business School.
63. Hayes, S. W., Priestley, J. L., Ishmakhametov, N., & Ray, H. E. (2020). "I'm not Working from Home, I'm Living at Work": Perceived Stress and Work-Related Burnout before and during COVID-19.
64. University of Illinois at Chicago, (2016). *Fear of the Unknown Common to Many Anxiety Disorders.* Science Daily.
65. Marr, B., (2021). *The Amazing Ways VR and AR Are Transforming the Travel Industry.* [Online]. Available: https://www.forbes.com (accessed on 07 September 2023).
66. Henderson, I., & Tsui, W., (2019). Chapter 17 – The role of niche aviation operations as tourist attractions. In: Anne, G. F. D., (ed.), *Contemporary Issues in Air Transport* (pp. 233–244). Elsevier.
67. Browaeys, M. J., & Price, R. (2008). *Understanding Cross-Cultural Management.* Pearson Education.
68. Ghosh, I., (2020). *Ranked: The 100 Most Spoken Languages Around the World.* [Online]. Available: https://www.visualcapitalist.com/ (accessed on 07 September 2023).

69. Hagiu, A., (2013). Strategic decisions for multisided platforms. *MIT Sloane Management Review.* p. Unknown.
70. Ramphal, R., (2017). Lean Six Sigma framework for the hospitality industry. *African Journal of Hospitality, Tourism and Leisure,* 2.
71. Michael, R. L., & George, L., (2002). *Lean Six Sigma: Combining Six Sigma with Lean Speed.* Texas: McGraw - Hill - Education.
72. Evangelia, F. V. C., (2019). The role of responsible leadership in creating a risk culture that promotes sustainable innovation in business. *Risk and Financial Management, 1*(1), Unknown.
73. Leon, J. P. D., (2019). *How to Spot Mental Health Issues and What to Do.* Gulf News Health.
74. Stuckey, H. L., (2015). The second step in data analysis is coding qualitative research data. *Journal of Social Health and Diabetes.* p. Unknown.
75. Laurel, T., (2019). *Humans, Robot Teams Work Better When There's an Emotional Connection.* https://news.umich.edu p. Unknown.
76. Brendan, (2018). *Why Hotels Should Use AB Testing to Drive Direct Bookings.* [Online]. Available: https://www.hotel-reservation-bot.de (accessed on 07 September 2023).
77. Mentalhelp.net, "Cognitive Theory of Depression – Aaron Beck" [Online] https://www.mentalhelp.net/depression/cognitive-theories/.
78. Cerasa, A., Fabbricatore, C., Ferraro, G., Pozzulo, R., Martino, I., & Liuzza, M. T., (2020). Work-related stress among chefs: A predictive model of health complaints. *Front Public Health,* 8–68.

CHAPTER 7

Artificial Intelligence in the Tourism and Hospitality Industry

ABHISHEK ACHARY and DIOGO DAVIDSON ALBUQUERQUE

Department of Humanities, Arts and Applied Sciences Amity University, Dubai, UAE

ABSTRACT

The word 'artificial intelligence (AI)' is coined by John McCarthy in 1956. AI has partially been used in many sectors like medical sector, industry sector, hospitality sector, business sector, that have aided the business opportunities, this had aided the economic growth. AI have also been used in hospitality sector like Alexa a voice commanded speaker system that can be activated through a voice, chat-robots that scans the ID of the guest and updates in the hotel server, that way the process of check-in and check-out can be quicker and service robots that does room service for the guest. AI has contributed to the global economy which gave opportunities for business. AI will be used very often in the upcoming years. The main objective of this chapter is to explore the different ways AI to be used in hospitality industry and how AI contributes to the industry.

7.1 INTRODUCTION

Artificial intelligence (AI) was first invented by John McCarthy in collaboration with Herbert A Simon, Allen Newel, Alan Turing, and Marvin Minsky. A conference was held in the summer of 1956 and John revealed the word AI to the world of data analytics [1]. Although robotics has been part of the technology for some time now, it has attracted attraction for its automation in recent times. We have since then evolutional in the robotics exceptionally under the Artificial Intelligent tag. Automation which can also be referred as robotics are used in multiple sectors in the healthcare industry to diagnose the illness and in the surgical domain and in Automobile industry mostly used in Cars, etc. They are also used in the knowledge sector, supply chain and more recently in the agriculture which benefits millions of farmers globally. Robots also have human collaboration where they are using for elderly assistance and even for the younger generation with special needs. Robots are being drastically used in the entertainment, civil force which helps in surveillance operations.

Robotics are also being used in the tourism and hospitality sector, where they are used in the front desk for customer interactions. Amazon's Robotic invention Alexa with its voice assistant feature along with Echo controlled speaker are also very popular in hospitality industry. Robotics is being researched further to see the level of usage and its acceptance amongst the human races. It has the potential to cross paths in every industry, and many have started exploring this option.

The hospitality industry has been inspired with the latest trends of technology and attempted to be trendsetter in many instances. The industry has multiple segments where the implementation of robotics is required such as Tourism, hotel, bar, front desk, industry kitchen, delivery, and service area. The usage of Robotics in the industry can boost the service standards not just due to the addition of a gimmick to attract guests however it would deduct the repetitive work for the staff and alter their focus toward the experience enhancement of the guests. Few hospitality sections are extremely dangerous and life-threatening, which may be supported by robots to avoid any human error, which at times leads to an accident.

The Robot Pepper [1] has been used for front desk solution helping with welcome, booking a table, making reservation, sending confirmations, calling a taxi and check out. The Moley's Chef Robots are equipped to cook dishes on the choice of guest [2]. The humanoid robots are providing the conversational flexibility and featuring seamless smart interactions between human and robots. Further, delivery and guiding and disinfection and Spot Robots are adding to the society of robots in Hospitality to provide service delivery in a restaurant, room or at the poolside. Whereas the guiding robots conveying directions and messages along with the disinfection.

While the exploration and usage of robotics are common in hospitality and tourism industry, the service industry is yet to analyze the potential of automation. AI are being used in innovative methods to understand the customer requirement based on the data gathered also suggest other relevance option under the same category, it is almost like a salesman providing his different offerings to the customers. The data analyzed can be used to predict the customer choice of food, etc. Robotics are being used for a contactless experience in the hotels where a guest can check-in and check-out with the human intervention by means of facial recognition, chat-bots, etc. [2].

7.2 ARTIFICIAL INTELLIGENCE (AI) IN GLOBAL HOSPITALITY INDUSTRY

Artificial intelligence (AI) as the potential to impact the gross domestic product (GDP) of a country's economy. Many countries have realized this potential and have started investing and is one of the most strategic investments in the recent times. Many corporations across the globe are trying a combination of labor force with AI technology to allocate the duties and roles equally. The research predicts that by 2030 the economic gains will reach by 45% which will be achieved by the product assortment, elegance, and affordances of a product over a period. China would be leading in terms of economic gains in the AI sector with 26% increase in their GDP by 2030 followed by North America with 14.5% which equals to USD 70 trillion and with a global impact of 70% financially [3].

The Hospitality industry has been estimated to grow fastest in the F&B segment, i.e., 55% of the share with notable gains in virtual assistants. The changes are predicted with a strongly build trust on devices during the period of Global pandemic due to rigged data analytics and customizable requirements. North America has deployed maximum utilization of the projects and hold 50% of the global market share [3]. The hospitality industry is utilizing multiple hotel operations, revenue management and analytics, personalized approach of customer service and marketing, and multilingual booking experience through natural language processing (NLP) [4].

The hospitality and Tourism industry is ever evolving, and the technological development has given boom to the industry always in the past.

The current study displayed the impact of AI on the industry around 400 billion dollars overall (Figure 7.1).

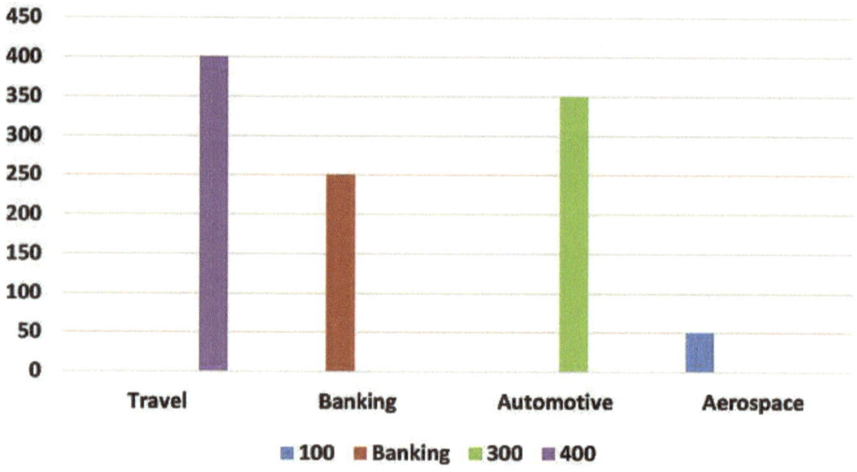

FIGURE 7.1 AI impacts derived from analytics (US$ in billions) [5].

7.3 UNITED ARAB EMIRATES (UAE): TOURISM AND HOSPITALITY SECTOR

As United Arab Emirates (UAE) is the world's most leading business sector and offers vast varieties of destination for tourists, its mission is to increase the awareness among the global audience and cause inward investment and inbound tourism inside every emirate. The hospitality sector of the country along with the tourism sector has formed itself as a sector which gives most of the employment in the state thus improving the economy by attracting funds. AI is playing a very important role in hospitality management in UAE. The plan is to transform UAE into a world leader in AI by investing people and industries that are key to their success. The hospitality sector in UAE is competitive. Technology can be a key weapon for the players in the industry for creating a commercial edge. The hospitality industry can be boosted by the use of AI robots that not only diminish the use of humans with the help of voice-activated assistants but also smarten the way hotel guest service is delivered to customer's satisfaction. AI-enabled robots or bots can offer minute service and careful assistance to the customers based on their personal needs. Tourism and hospitality is a highly visible and successful export sector of UAE. AI in hospitality will help the business owners to deliver more tailored experience than ever before, analyze users reviews and offers virtual assistance. UAE is still boosting its hospitality sector by making use of AI and attracting more customers. Many hotels in UAE have adopted the use of AI and technology which helps the customers to plan their stay before their trip. The ease in visiting UAE without any questions after coming as the customers already plan their stay is making hospitality sector better. The revenues are also increased.

7.4 EXPO 2020: IMPACT IN TOURISM AND HOSPITALITY SECTOR

Expo 2020 Dubai has a profound symbolic value because it's a first kind of event in the Middle East, North Africa and South Asia. The theme of Expo 2020 was connecting minds and creating the future. This event was created to promote communication and interaction and open up better opportunities for trade and investment exchange between all countries. The real estate, tourism, and hospitality sector are the major recipients of foreign and domestic investments attracted by Expo 2020. It would be entirely reasonable that Expo 2020 Dubai explored the world of AI. Expo 2020 was the first commercial customer that accessed 5G in the region and they built a smart site that not only created exceptional visitors attractions and enhanced experiences on digital platforms. AI was the very much part of the project. They used AI to enhance visitors experience, create shorter and faster queues, provided more personalized access to information and ensured security of the event. Visitors were able to access all the information from the chatbots and asked questions. Expo 2020 also played a significant role in the improvement of hospitality industries. Around 25 million and more visitors were expected to visit the Expo. The Dubai hospitality industry attracted exceptional growth, with enough rooms joining the market. Due to this remarkable growth, many hospitality jobs were available in a dramatic way; there was massive opportunity for young talents. Expo provided a unique and international stage to engage in cultural diplomacy and interact with the host country, participants, potential tourists, trade partners and investor. Expo 2020 Dubai's three themes of opportunity, mobility, and sustainability inspired the global visitors to preserve and protect the planet, explore new frontiers and build better future. The F&B offered had a culinary paradise. The hospitality and tourism received a huge push through the event. It greatly contributed to the economy as well as sustained progress of this country.

7.5 POPULAR EXAMPLES IN UAE: TECHNOLOGY AND HOSPITALITY SECTOR

The four emerging hotel technologies can be defined as Hotel CRM, virtual, and augmented reality (AR), revenue management tools and contactless check-in (Figure 7.2).

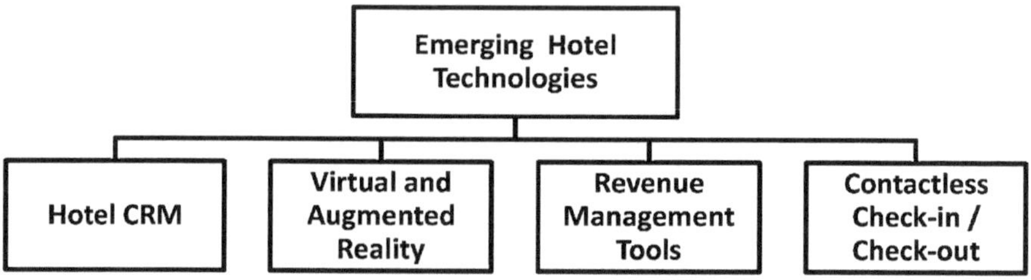

FIGURE 7.2 Emerging hotel technologies.

1. **Hotel Customer Relationship Management (CRM):** It can be very useful for hotels take full advantage of customer's history, loyalty conversations, as well as know more behavioral aspects of customers during their stay. An exceptional experience can be delivered through personalization by consolidating all guest data into a database in hotel CRM system.
2. **Hotel Virtual and Augmented Reality (AR):** It is also gaining traction. There are many VR experiences that tourist can try, from example the largest VR park to Dreamscape in the Mall of Emirates in Dubai. AR hospitality technologies allow customers to see and feel for what they expect when they stay at a hotel even before they are physically on site. AR can offer interactive hotel rooms and gamification. Rove Hotels which launched the first AR game in Dubai and Burj Al Arab's impressive tour which incorporates

AR to create more impressive tour experience are two main examples.

3. **Revenue Management:** Hotels also rely on revenue management tools to accurately price their rooms. It takes into account factors such as location, occupancy rates, and historical data, although it is not a new technology in the hospitality industry, revenue management practices took more of a rule-based approach to price decisions.
4. **Contactless Check-In:** In recent years, hotels with contactless check-in have gain popularity in the world. The customers love this because of the reduction of queues at front desk and they can perform check in and check out by themselves. Almost every hotel in UAE is using this by adopting AI and attracting lots of guests from around the world. Every hotel supports RFID hotel key card, door locking system.
5. **Atlantis Group of Hotels, in Dubai:** It has entered into an agreement with Digital Ecosystem to develop an AI to capture guest's feedback on indoor sports and recommend the appropriate sports based on their likes and choice to their guests depending on the event availability. They have also implemented the blockchain technology with which a customer is provided with a wrist band which acts as their digital wallet, where they can charge them and use it in the in-house services offered during their stay in the hotel [4].
6. **Shangri-La Group of Hotels, Abu Dhabi:** This which has a massive room's capacity, has successfully implemented a unique Guest Operation Automation System. This system is developed to automate post check-in services which means when a room reservation is termed as Occupied, it switches on the air-cooling system in the room to a moderate temperature while the visitor heads to their room making the guests feel relaxed and which then turns a customer satisfaction. It also provides the guests with a guest response time data, statistics on energy saving and popularity of the hotel by providing the occupancy of the hotel which provides a total transparency to the guests [5].

7. **Emaar Entertainment, Dubai:** This have opened a VR park famously known as VR Gardens, which is a mixture of augmented and VR. Its unique rides fun-filled games, and thrilling adventures are built upon this unique concept. They are also constantly in lookout of the startups to explore further in this sector and enhance their customer experience to another level. One of the recent addictions is to showcase the historic past, amazing present and an unimaginable future of Dubai using this concept. They have chosen Winnow to deliver another remarkable concept to reduce the kitchen waste with their AI tool of Winnow Vision, across the emirates. This will basically help to track meals leftovers. It also practices laptop visualization to assist cooks effortlessly in pinpointing food left-over, storage time which in turn helps to cut cost [6].

7.6 ARTIFICIAL INTELLIGENCE (AI) APPLICATION IN HOSPITALITY AND TOURISM

The revolution of AI and robots has provided tourism organizations and businesses with new opportunities. AI is a modern form of intelligence which has the ability to synthesize several different ideas simultaneously. Modern travelers expecting immediate responses in online platforms. However, chat-bots and tour bots enable personalization service for tourism, and also aid in productivity and which aids in the economic growth [7].

AI is a new form of intelligence. Today technological innovation requires efficiency, sustainability, and productiveness at the same time. AI could be used in several specific fields which are increasing in commercial sectors. The performance of interactive communication is a key importance for innovation, companies to increase the sales of growth. The chat-bots and other clever solutions which are used by AI provider allow the development of customized offers. Although in the tourism sector AI-related responses are still not substantial they're predicting that AI will be used in the future growth and enhance the services to the guest [8].

7.7 TOURISM AND HOSPITALITY INDUSTRY: IMPACT OF AI

Understanding private preferences and tour histories is fundamental to offering Customized products and offerings. New hospitality business models have changed the way customers see the resort industry and caused higher price sensitivity. Therefore, retaining consumer loyalty has end up being superior. Global Data's 2016 show that 47% of global respondents bear in mind loyalty applications essential when booking a hotel for his or her holidays. Names which include Hyatt, Marriott, Best Western, and Hilton have already attempted to capitalize on this fashion with the aid of establishing guest reward packages to attract customers to their provider and, more importantly, to acquire valuable information. In a few cases, system studying analyzes these personal statistics to build character person profiles and makes clever recommendations, offering services that better suit consumer needs [9].

Big chains have implemented virtual voice assistants inclusive of Amazon's Alexa or Google Home to decorate customer revel. Hilton has gone further, by imposing Connie, a robotic concierge primarily based on IBM Watson's device mastering algorithm, in certainly one of its accommodations in Virginia Connie is based totally on a cognitive intelligence that learns, adapts, and improves its tips the more it interacts with customers Starwood motels, in cooperation with robotics employer Sacioke, has also trialed two robot concierges called A.L.O. in its Cupertino Aloft inn. In Japan, Henn-na Hotel has been replaced all provider assistants with robots. Increase in the AI in hospitality sectors could lead to the decrease in the number of human workers which fears many workers and also drop down of jobs [10].

The technology used in Hilton is chat robots the guest enters the hotel and through the eye scanner the robot scans the id of the guest and it updates the data base of the hotel server and also this way it enhances the guest experience and also can increase the revenue of the hotel the robots also understand the guest expression and react accordingly to manage the guest, also builds the character profile and makes clever decisions in guest service.

Hilton has long gone further, by enforcing Connie, a robotic concierge based on IBM Watson's device getting to know algorithm, in certainly one in all its motels in Virginia Connie is based, totally on a cognitive intelligence that learns, adapts, and improves its recommendations the more it interacts with customers Starwood motels, in cooperation with robotics business enterprise Sacioke, has also trialed two robot concierges known as A.L.O. in its Cupertino Aloft inn. In Japan, Henn-na Hotel has even modified all of its company assistants with robots [10, 11].

7.8 FUTURE OPPORTUNITIES AND IMPLEMENTATION

The usage of AI has helped in many sectors like engineering sector, hospitality sector [12, 13], business sector [14, 15], industry sector. The usage of AI can easily be seen nowadays in processing the patient information, service robots, chat-bots, self-driving vehicles, robots in the production line. In the upcoming years there will be a rapid boom of AI usage [16]. Enterprises believe making an investment in AI will result in greater aggressive advancement. AI will open up new agencies in the market which will bring up the business opportunities.

As businesses continue to use AI technologies within their operations, they are gaining many benefits including multiple gains. The revenue generated by AI increased over time in many of the functions of business by using AI technologies. Efficiency and productivity gains are the two most important benefits of implementing AI into the enterprise. The technology handles the tasks at pace and scale that humans are not able to match. AI allows workers to move to a higher level in performing the tasks that technology can't do. This allows the businesses to minimize the cost. As fast as business moves in this digital era, AI helps it to move even faster. AI enables shorter development cycles and shortens the time it takes to move from design to commercialization and this in turn provides better and more immediate return on investment on development dollars. Executives can also use AI in the expansion of their business model said by Chris Brahm, a leader at global advanced analytics practice. As executives deploy data in their businesses, it opens opportunities for more businesses to participate in different areas. Improved customer service is also one of the benefits for businesses by adopting AI. AI can lead to more customized and personalized interactions between organizations and customer. It helps the businesses to make faster decisions by using certain technologies. AI's power to store and process large amount of data means that business can implement monitoring capabilities which will alert them to the problems and issues recommend action and, in some cases, gives a response. AI's monitoring ability can be used in other areas as well such as cyber security operations where large amount of data needs to be analyzed and understood. It also avoids mistake and human error and grows the enterprise by enabling analysis and giving advice and support. Businesses are using AI in improving several areas of talent management from streamlining the hiring process to pointing out bias in corporate areas. Enterprises believe in investing in AI which will lead to greater competitive advantages. AI will open up new agencies as well as bring up competition, new methods to the markets. This can aid in the economic growth globally. The usage of AI also brings innovative ideas in the market.

Current users of AI believes that AI can determine the income forecasting and for improving email marketing. Companies is planning to use AI for income forecasting. This can help the companies to understand the income status and plan for the next move. The above graphic compares adoption prices of present-day AI versus all respondents.

7.9 CONCLUSION

John McCarthy is the "founding fathers" of AI, together with Alan Turing, Marvin Minsky, Allen Newell, and Herbert A. Simon. McCarthy coined the term "artificial intelligence" in 1955, Artificial intelligence or robots are coming into travel, tourism, and hospitality. Wynn inn in Las Vegas introduced Amazon's Echo voice-controlled speaker. Artificial intelligence and robotics provides vast opportunities to travel, tourism, and hospitality sectors to improve their operations, productivity, deliver steady product and transfer the same service delivery method to the customers.

AI based meals waste technology lets in kitchens to automatically track meals waste. Many hotels have also used chat robots that is when the guest enters the hotel the chat robots scan the ID card of the guest and updates the details in the hotels data base this way the process of check-in and check-out will be faster and also through voice commands like Alexa, Siri, etc., there are various pros and cons to this concept. Some of the best benefits we can quote can be implementation of AI can provide personalized service, can help in increase in the revenue of the hotel and the economy of the country which also contributes on global economy. But at the same time, implementation of AI also comes with few challenges such as, initial set up cost can be very high, regular maintenance is needed, and increase in intensive technologies can lead to decrease in the human employment in few industries.

KEYWORDS

- **AI applications**
- **artificial intelligence**
- **economic growth**
- **global advanced analytics practice**
- **hospitality sector**
- **intensive technologies**
- **tourism**

REFERENCES

1. Moor, J., (2006). The Dartmouth college artificial intelligence conference: The next fifty years. *AI Magazine*, *27*(4), 87.
2. Ivanov, S. H., Webster, C., & Berezina, K. (2017). Adoption of robots and service automation by tourism and hospitality companies. *Revista Turismo & Desenvolvimento*, *27*(28), 1501–1517.
3. Walsh, T., (2020). *A Pebble in the AI Race.* arXiv preprint arXiv:2003.13861.
4. Andrea, B. *Digital Transformation at Atlantis the Palm Points to Future of Hospitality.* [Online] https://www.cio.com/article/189144/digital-transformation-at-atlantis-the-palm-points-to-future-of-hospitality.html (accessed on 07 September 2023).
5. Al-shami, S. A. H., Al Mamun, A., Ahmed, E. M., & Rashid, N., (2021). Artificial intelligent towards hotels' competitive advantage. An exploratory study from the UAE. *foresight.*
6. Popesku, J., (2019). Current applications of artificial intelligence in tourism and hospitality. In: *Sinteza 2019-International Scientific Conference on Information Technology and Data Related Research* (pp. 84–90). Singidunum University.
7. Zsarnoczky, M., (2017). How does artificial intelligence affect the tourism industry? *VADYBA*, *31*(2), 85–90.
8. *Nielsen-Global-Retail-Loyalty-Sentiment-Report*, [Online] https://www.scribd.com/document/355244160/Nielsen-Global-Retail-Loyalty-Report-f-1 (accessed on 07 September 2023).
9. Martinez-Torres, M. D. R., & Toral, S. L., (2019). A machine learning approach for the identification of the deceptive reviews in the hospitality sector using unique attributes and sentiment orientation. *Tourism Management*, *75*, 393–403.
10. James, W. *Aloft Rolls Out Robotic Butler.* [Online] https://www.hotelmanagement.com.au/2014/08/14/aloft-rolls-robotic-butler/#:~:text=Aloft%20Hotels%20has%20unveiled%20A.L.O.,In%20this%20role%2C%20A.L.O (accessed on 07 September 2023).

11. Esther, H. *Japanese Hotel's Robots Modified to Prevent Security Risk.* [Online] https://www.hotelmanagement.net/tech/japanese-hotel-bedside-robots-hacked (accessed on 07 September 2023).
12. Grobbelaar, W., Verma, A., & Shukla, V. K., (2021). Analyzing human-robotic interaction in the food industry. In: *Journal of Physics: Conference Series* (Vol. 1714, No. 1, p. 012032). IOP Publishing.
13. Mohammed, U. S., Shukla, V. K., Sharma, R., & Verma, A., (2022). Next step to the future of restaurants through artificial intelligence and facial recognition. In: *Cyber Intelligence and Information Retrieval* (pp. 89–101). Springer, Singapore.
14. Suhel, S. F., Shukla, V. K., Vyas, S., & Mishra, V. P., (2020). Conversation to automation in banking through chatbot using artificial machine intelligence language. In: *2020 8th International Conference on Reliability, Infocom Technologies and Optimization (Trends and Future Directions) (ICRITO)* (pp. 611–618). IEEE.
15. Bala, S., Khalid, M. N., Kumar, H., & Shukla, V. K., (2022). The practical enactment of robotics and artificial intelligence technologies in e-commerce. In: *Cyber Intelligence and Information Retrieval* (pp. 455–467). Springer, Singapore.
16. Petiwala, F. F., Shukla, V. K., & Vyas, S., (2021). IBM Watson: Redefining artificial intelligence through cognitive computing. In: *Proceedings of International Conference on Machine Intelligence and Data Science Applications* (pp. 173–185). Springer, Singapore.

CHAPTER 8

Automation and Robotics of the Tourism Industry

MUHAMMET ERBAY

Faculty of Tourism, Selcuk University, Turkey

ABSTRACT

The Tourism industry has impacted several lives across the world and has impacted the economic status of the countries. The technology has transformed it multifold in less than expected time. The current era is in the pursuit of an artificially intelligent Tourism industry. The industry has already started adapting to the future of travel with its leaps and bounds. However, automation has brought challenges with Robotics and its day-to-day implementation in the tourism industry. The chapter focuses on the industry's growth with automation, Robotics, its impacts, challenges, and solutions. The chapter predicts the future of robotics and automation in the tourism industry.

8.1 INTRODUCTION

Artificial intelligence (AI) is among the most brilliant discoveries that have transformed numerous sectors worldwide in today's technologically advanced world. The development of digital platforms to collect intellectual activity can be termed AI [1]. In the Dartmouth Summer Research Project of John McCarthy, AI first emerged in 1956. Over the years, various AI victories have taken the form of heuristic research, characteristic identification, facial identification systems, NLP and the idea of mobile robots. In the 1980s, substantial fundamental technological development was accomplished, and submission has improved dramatically [2]. The technology environment made significant progress in the 1990s, especially in AI. The evolution was primarily due to sophisticated and innovative technology that enabled technologists to utilize vast information capacities, create robots efficiently and strive towards greater processing power. Ever since the AI notion has come to a distance and revolutionized technological management, AI finally culminated in the 21st century and can transform individuals, enterprises. AI has infiltrated the tourism and hospitality industries, employing it to obtain competitiveness in a growing environment.

Technology has gradually found its way into the tourism industry in Uzbekistan, reducing the number of human operations. As a long-standing early adopter of technology, the tourism industry has embraced AI with great enthusiasm. Tourism, like every other business, is facing an increasingly automated future. Tourism providers have begun to include intelligent machines in their operations. Travel buddy, a robotic suitcase, is one of the most forward-thinking uses of robotics in today's travel business. The suitcase's collision detection and 360° turning capabilities make it possible to eliminate the requirement for a person to carry a case. From Nagasaki, Japan, comes one of the most spectacular robots in the tourist and travel sector. The Henn-na Hotel is the world's first robot-staffed hotel, with robots serving as front desk agents, client data points, and storage. The robots use technologies such as speech recognition, facial recognition, and AI. In 2018, aviation was developed and launched a self-driving trolley called Care-e to help fly passengers carry their bags to their airport, following the debut of an androgyne robot, Spencer, to support airport passengers [3]. Japanese adventure park Huis 10 Bosch constructed Henn Hotel, the first robot hotel globally, in 2015 [4]. The market share has now been enlarged to several other venues. The hotels employ dinosaur receptionists, robotic carriers, robotic wardrobes and in-room administrative help to mention just a few. While an automatic hotel is still rare today, hotels globally have adopted customer-facing smart automation, including four independent verification systems, virtual administrative assistants and hotel robots [5]. Tourism researchers have begun to focus on AI, robot technology and automation. Their contributions are nonetheless restricted to applications and potential future deployments and consequences.

8.2 LITERATURE REVIEW

Following the industrial revolution, technological advancements made their way into the services business, presenting occasions for service robotics [6]. Automation is a term that refers to the process of using machines to carry out a "predetermined or reprogrammable work sequence." Automatic teller machines, elevators, self-checkout businesses, and vending machines were early examples of automated procedures. ICT advancements contribute to the ongoing improvement of the client experience and service productivity [7]. Self-service check kiosks, for example, may let tourists spend less time waiting at the airport, move faster to their gate, and improve their customer experience.

Similarly, smartphone applications such as No Wait allows visitors to view wait times at nearby restaurants. The application allows users to add themselves to the waiting list, monitor how many groups are ahead of them, and arrive at the restaurant at the appropriate time to sit down, ensuring a flawless eating experience [8]. Other technologies that can help automate services involve, but are not confined to, 3D modeling, self-driving autos, and robotics. Robots can be thought of as "intelligent physical devices" [9] that possess a level of discretion, mobility, and sensory capability to perform predefined tasks [10]. This autonomy may be predisposed by the difficulty of the setting in which a robot works and the robot's inherent characteristics such as intelligence, flexibility, and sensory capabilities. Sensors are integrated plans that enable a robot to interact with and learn about its environment. The primary activities of a robot typically require the use of sensors. These sensors commonly simulate human senses, including light, pressure, taste, and hearing [11]. Industrial machines are employed for various industrial responsibilities, including welding, pelletizing, and other manufacturing and production tasks [12].

However, services robots are scheduled for physical and social contact to help and serve people. Service robots can also be classified as specialized services and personal services. Industrial robots and service robots continue to expand to the International Robotics Federation. For example, sales for industrial robots rose by 0.5% in the pandemic conditions, with 383,545 units in the year 2020. The operational stock of the industrial robot was computed 3,014,879, which records a 10% increase this year [10]. The hotel and tourism industry attracted robots as a service industry.

8.3 HISTORY AND EVOLUTION OF TOURISM, AUTOMATION, AND ROBOTICS

One of the most important is AI, which is the brilliant discovery that has transformed numerous sectors worldwide in today's technologically advanced world. AI is the process of creating computer systems capable of doing jobs and activities that need human intellect [1]. In 1956, John McCarthy presented the idea of the IoT in the research project. IA breakthroughs have taken place over the years, including systematic searches, identification of attributes, biometric identification systems, natural language translation, and mobile robots. There were enormous innovations in the 1980s, and their use rose significantly [2]. As technology advanced in the 1990s, AI became a significant focus. The increased population mainly resulted from advanced and new technologies that allowed experts to efficiently process vast amounts of information, develop robots, and engage in enhanced computer capabilities. Since then, the AI paradigm has advanced significantly, transforming the capabilities of technology. AI has crossed a threshold in the 21st century, and it can have a tremendous influence on humans, businesses, and industries. In the modern digital era, AI technology is employed in many companies, not simply the information technology (IT) business [13]. For example, AI can be found in self-driving cars, automated nurses, tracking systems, chatbots, and human vs machine video games.

Each generation has experienced technological advancements that have transformed the tourism business. By the mid-nineteenth century, the train had made it possible for tourist groups to travel together, and the first planned tours had begun. The automobile facilitated travel worldwide in the early 20th century by allowing tourists to plan more flexible itineraries. Jet planes revolutionized long-distance travel for the working middle class in the latter half of the 20th century. Online Services, the Internet, social media, websites, and mobile applications shrunk the world at the beginning of the 21st and early 21st century. They revolutionized the travel, tourist, and hospitality industries [14]. An even greater and more far-reaching uprising has struck the tourism industry recently. Automated travel will be the norm in tourism soon. Tourists' supply and demand sides will be affected by new realities, which significantly impacted the previous tourism revolutions. Robotics in tourism necessitates a shift in perspective from the assumption that a hospitality professional or customer is human. Robots could provide tourism and hospitality services in the future.

8.4 USES OF AUTOMATION AND ROBOTICS IN TOURISM

Table 8.1 highlights several prominent service automation and robot implementation examples in the tourism and hospitality industries.

8.5 FINDINGS AND DISCUSSION

8.5.1 THE CURRENT AUTOMATION AND ROBOTICS FEATURES IN THE TOURISM INDUSTRY

In the tourism industry, various innovative technologies have emerged in AI. These technologies have helped provide tourists with a unique experience. These technologies include biometrics, increased reality, chatbots, Google Maps robots, language travel, audio tours, and shopping facilities. Biometric technology is an application of AI technology that acquires traction and application in numerous fields for various purposes. In the tourist and tourism industries, facial recognition is very widely adopted. Travel documents must be verified by official institutions such as taxes, immigration, and airports, and this can be a time-consuming process for tourists. A further AI being used in the tourism industry is robots. Simple chores such as switching on the bedroom light, disconnecting the TV, managing base management systems to ensure the check-in is automated and welcoming hotel visitors are performed by technological aids using IoT (internet of things), technology. Robot receptionists have grown widespread in tourism, a significant impact on the way consumers and guests relate to their establishment and experience it. When customers arrive at a hotel for the first time, housekeeping is also their responsibility. Nagasaki, Japan, is home to one of the most impressive robots in the travel and tourism business. The Henn-na Place is considered the world's first automation hotel because of its front desk agents, customer data points, and storage robots. The robots utilize voice, facial recognition, and AI technology.

Language translators are a small number of software tools that translate between languages. These programs are especially beneficial for tourists

TABLE 8.1 Examples of Service and Automation in Tourism and Hospitality

		Presently in Use	A Possible Application
Hotels	Service Automation	• Self-service check-in kiosks. • Check-in through a mobile device. • Mobile service requests.	• Automation as a service
	Robots	• Robots at the front desk. • Delivery robots. • Room assistant robots. • Robotic vacuum cleaners.	• Hospitality robots, for example, cleaning, washing. • Robot luggage porters and butlers are being developed to navigate hotels and provide valuable services to the client, airport security robots.
Airports	Service Automation	• Bag check. • Conveyor belts. • Check-in kiosks for self-service. • Gates with self-boarding	• Full-service automation.
	Robots	• Airport self-guide robot. • Robots of the bag drop. • Entertainment's robots. • Cleaning robot. • Delivery robot. • Robotic agents for customer service.	• Robotic carts for baggage. • Cars for self-driving passengers between gates.
Tourist information agencies and centers	Service Automation	• Displays and information kiosks. • Audio guides.	• Information, booking, and payment self-service kiosks.
	Robots	None	• Robots' guides. • Robotic sales representative. • Robotic digital assistant for the home that is connected to the website of a tour operators.

visiting distant countries and encountering strange languages. Traveling to a remote nation can be challenging, even more so when linguistic barriers exist. Google Maps, which uses location tracking, aided travelers by providing instructions. Google Maps' AI technology increased the information by informing tourists about traffic crashes bottlenecks. While Google Maps facilitates travel by discovering the correct routes, tourists frequently experience a similar issue. Customers finish their journey by traveling entirely differently, which is found after roaming a world away [15]. A growing number of airports rely on robots to assist human security officers in keeping passengers safe. Detecting weapons and other forbidden items in carry-on luggage is a common application for Knightscope robots, which may be found in various airports.

8.5.2 THE EFFECT OF AUTOMATION AND ROBOTICS IN THE TOURISM INDUSTRY

Chatbots are generally automated computer programs that are pre-programmed to reply to simple explorations posed by customers [16]. Customers enter hotel rooms with many queries about the destination's tourism structure and tourist substructure. Customers ask several inquiries to the chatbots about the hotel's facilities, lunch and dinner hours, menu items provided during lunch and dinner, gym services, hotel amenities, and security measures. Robots surprise clients by giving unanticipated activities, which keeps them interested and creates a fresh and excellent contribution in their minds. Helping tourists in the hotel includes directing them to their rooms, delivering luggage to their rooms, completing house-keeping responsibilities, and serving meals and beverages [17]. Thus, robots continually boost client happiness and engagement in hospitality facilities by providing new and creative services.

Using Location-based data, Google Maps' AI provides information on highways, public transportation, and private automobiles. Nowadays, GPS-based AI in cars can offer information about road conditions and traffic bottlenecks and arrange an alternate route to avoid them. Information on surrounding stores and businesses, hotels, and malls, cinemas, and restaurants will be included in the next edition of Virtual Server to allow for destination experiences. Passengers at airports can check-in using facial recognition technology, which can eliminate the need for immigration, customs, and other government agencies to verify their identity. In addition to speeding up check-ins, this facial recognition system uses blockchain technology to ensure travelers' personal information [18]. Customers benefit from these innovations since they make travel more convenient.

Two years ago, several high-profile hotel chains rolled out "Smart Rooms" that integrated voice assistants such as Alexa and the Amazon Echo. It is possible to use these voice-activated digital butlers to perform several activities, from identifying nearby attractions to turning down the lights, thanks to the IoT. Due to the COVID-19 pandemic's requirement for frictionless service, the benefits of an Automated consultant became even more apparent. This cutting-edge technology can not only provide customers with an exciting and secure service, but it can also record and process crucial data. In the wake of the economic recovery, hotels are now looking into how AI assistants may help them stay on top of the competition.

Since its inception, the concept of "when your phone becomes your tool for practically everything," or "mobile application integration," has been gaining traction. The COVID-19 epidemic has propelled adoption to an unprecedented level. Many countries have seen an increase in meal delivery services like Menulog, UberEATS, and Deliveroo since the outbreak of the epidemic. As a result, restaurants have developed contactless menus and mobile ordering, which allow consumers to order by swiping their smartphones rather than reading a food label.

8.6 CHALLENGES AND SOLUTIONS FOR AUTOMATION AND ROBOTICS

Although Robotics is advancing quickly, there is still a barrier to its use by illiterate people [19]. These limits have to be addressed and are combined with new solutions, technological development and updated policies. Regardless of what sectors AI replaces human labor and argues if AI will eventually replace people's intelligence, the objective is to address various unsolved concerns across many company segments and resolve them. The safety and security of the database is another critical problem in the deployment of AI. This is a chronic difficulty with AI in several essential areas, including banking and defense. For example, while Blockchain technologies are integrated with Facial Recognition software, providing information security and security, certain governments remain reluctant to use technology because of privacy and safety concerns [20]. Chatbots and hotels maintain histories of purchases and holidays, creating worries about privacy and data protection. It represents a security and privacy threat [21].

When even a minor viral attack compromises an internet service, it might cause even more instability in these application services. As a result of the substantial investment required for small service organizations for AI technology, it is not feasible [10]. Even though computers and robots can swap human employees, machines, such as chatbots, are confined to responding to simple questions. These technologies provide replies based on the keywords contained in the queries. When a critical or complex issue requires resolution, customers rely on manual effort.

8.7 THE FUTURE OF AUTOMATION AND ROBOTICS

Numerous reports forecast a massive growth in the forthcoming number of robots. Several of these will be robotics soon. Meanwhile, robots and autonomous systems are likely to become increasingly prevalent in the future society, especially self-driving vehicles and service robots for work and hotels. The technology that surrounds us comes in various procedures and has variable degrees of expansion and influence on our lives. AI enables technology to adapt via learning to allow systems to sense, reason, and behave optimally [22]. AI has seen significant development in recent years in various corporate domains, including customer service and decision assistance in the tourism industry. A *service robot* is a robot that runs entirely independently to perform valuable activities for humans but does not include industrial automation applications. They are now used in a limited number of settings, such as internal airport transit.

The technological progress from industrial robots to service robots reflects a shift toward more customized systems with great independence. In contrast, the need for industrial robots has been there for years. But at the same time, the influence of service automation in the office and at home has yet to be determined. Advances in AI explore will significantly influence the rate at which intelligent and autonomous service robots become available. Section "When and Where Will the Big Revolution Come?" discusses several aspects that could contribute to this technological advancement. Blockchain, AI, big data, smart mobility, and renewable energy are just a few of the new technologies that have been embraced in the construction of innovative, sustainable cities by the federal government's policymakers. The national level must support and synchronize its strategy for a smart, sustainable city to thrive.

8.8 CONCLUSION AND RECOMMENDATION

The growth of the AI mark in the tourism business is a promising indicator in the technology-driven moment. It shows that the sector can optimize its operational efficiency with modern innovations. Tourism sector clients in Uzbekistan will benefit from an improved satisfaction level. At the same time, tourism operators can have more influence over the procedures. The business operations will be fully automated and optimized for operational activities and standards. An unclear future awaits AI in the tourism industry. However, there is reason to be hopeful.

Society can deal with some of the most critical concerns that AI presents. Employees and Automation technologies will be able to converse, and problems about privacy and connectivity will be solved to allow AI systems to be deployed. Intelligence, in this view, can be viewed as a collection of technology aimed at increasing the overall enjoyment of tourists. As a result, corporations will better serve their customers by providing customized products, services, and experiences based on their unique demands. Customized packages for clients will be possible in the future, as well. As a result, the total cost of operations will be reduced, and some of the savings will be passed on to customers in part. Due to this, businesses will be able to provide services that were previously out of reach at a more accessible price. Additionally, AR may support staff to multitask by assigning the repeated task to its virtual assistants [23]. As a result, customer service and satisfaction can be improved by using technology to supplement specialized professions or relieve humans of specific duties.

Customers perceive AI as allowing them to plan their holidays more quickly, at reduced transaction costs, and with a package tailored to their specific preferences and interests. As a result, they will obtain prognostic offers that are specifically customized to their needs. Passengers will be aided by technology throughout the journey, which will alleviate their fear and anxiety of the unknown. Traveling will no longer be hampered by language and cultural barriers but enhanced by them. Customers will receive the best service possible while retaining the highest possible level of privacy [24].

KEYWORDS

- **artificial intelligence**
- **automation**
- **robotics features**
- **robotics tourism**
- **service robot**
- **tourism industry**
- **virtual assistants**

REFERENCES

1. Benckendorff, P. J., Sheldon, P. J., & Fesenmaier, D. R., (2014). *The Digital Tourism Landscape Tourism Information Technology* (pp. 22–52).
2. Bowyer, K. W., (2004). Face recognition technology: Security versus privacy. *IEEE Technol. Soc. Mag., 23*(1), 9–19.
3. Chen, Y., & Hu, H., (2013). Internet of intelligent things and robots as a service. *Simulation Modelling Practice and Theory, 34*, 159–171.
4. Colestock, H., (2005). *Industrial Robotics: Selection, Design, and Maintenance*. New York: McGraw-Hill.
5. Collier, D. A., (1983). The service sector revolution: The automation of services. *Long Range Plann., 16*(6), 10–20.
6. *Executive Summary World Robotics 2016 Service Robots*. Ifr.org. [Online]. Available: https://ifr.org/downloads/press/02_2016/Executive_Summary_Service_Robots_2016.pdf (accessed on 07 September 2023).
7. Issa, H., Sun, T., & Vasarhelyi, M. A., (2016). Research ideas for artificial intelligence in auditing: The formalization of audit and workforce supplementation. *J. Emerg. Technol. Acc., 13*(2), 1–20.
8. Ivanov, S. H., & Webster, C., (2017). Adoption of robots, artificial intelligence, and service automation by travel, tourism, and hospitality companies-a cost-benefit analysis. Artificial intelligence and service automation by travel. In: *Tourism and Hospitality Companies-A Cost-Benefit Analysis*.

9. Kannan, P. V., & Bernoff, J., (2019). The future of customer service is AI-human collaboration. *MIT Sloan Management Review.*
10. Law, R., Buhalis, D., & Cobanoglu, C., (2014). Progress on information and communication technologies in hospitality and tourism. *Int. J. Contemp. Hosp. Manag., 26*(5), 727–750.
11. Lommatzsch, A., (2018). A next-generation chatbot framework for public administration. In: *International Conference on Innovations for Community Services* (pp. 127–141). Cham: Springer.
12. Marouane, C., Maier, M., Feld, S., & Werner, M., (2014). Visual positioning systems-an extension to MoVIPS. In: *2014 International Conference on Indoor Positioning and Indoor Navigation (IPIN)* (pp. 95–104). IEEE.
13. Murphy, J., Hofacker, C., & Gretzel, U., (2017). The dawning of the age of robots in hospitality and tourism: Challenges for teaching and research. *European Journal of Tourism Research, 15*, pp. 104–111.
14. Oh, K. J., Lee, D., Ko, B., & Choi, H. J., (2017). A chatbot for psychiatric counseling in mental healthcare service based on emotional dialogue analysis and sentence generation. In: *2017 18th IEEE International Conference on Mobile Data Management (MDM).*
15. Techcrunch.com. [Online]. Available: https://techcrunch.com/2014/01/30/nowait-the-app-that-lets-you-join-a-restaurant-wait-list-from-your-phone-goes-nationwide/ (accessed on 07 September 2023).
16. Rajesh, M., (2015). *Inside Japan's First Robot-Staffed Hotel.* The Guardian.
17. Reddy, R., (2006). Robotics and intelligent systems in support of society. *IEEE Intell. Syst., 21*(3), 24–31.
18. Ruocco, S., (2013). *Robot Sensors and Transducers.* Springer Science & Business Media.
19. Russell, S., & Norvig, P., (2016). *Artificial Intelligence: A Modern Approach.* Pearson, New York, NY.
20. Torresen, J., (2018). A review of future and ethical perspectives of robotics and AI. *Front. Robot. AI, 4.*
21. Tussyadiah, I., (2020). A review of research into automation in tourism: Launching the annals of tourism research curated collection on artificial intelligence and robotics in tourism. *Ann. Tour. Res., 81*(102883), 102883.
22. Patel, V. *Airport Passenger Processing Technology: A Biometric Airport Airport Passenger Processing Technology Journey* (Doctoral dissertation, Thesis. Embry-Riddle Aeronautical University. Florida. 2018. 80 p. URL: https://commons.erau.edu/cgi/viewcontent.cgi).
23. Nagaraj, S., (2020). Marketing analytics for customer engagement: A viewpoint. *International Journal of Information Systems and Social Change (IJISSC), 11*(2), 41–55.
24. *How to Access Research Remotely.* Cabdirect.org. [Online]. Available: https://www.cabdirect.org/cabdirect/abstract/20193008127 (accessed on 07 September 2023).

CHAPTER 9

Role of IoT in the Tourism Industry

DARRYL MATHEW,[1] VINOD KUMAR SHUKLA,[1] FATIMA BEENA,[2] NITIN SHANKAR,[3] and SANA MOID[3]

[1]*Department of Engineering and Architecture, Amity University, Dubai, UAE*

[2]*De Montfort University, Dubai, United Arab Emirates*

[3]*Amity Business School, Amity University, Lucknow Campus, Uttar Pradesh, India*

ABSTRACT

The Internet of Things (IoT) is a concept connected to the internet, which works with sensors. The main objective of this technology is to process the data from these sensors. Today, there are billions of smart devices around the world connected to the internet and helping us in our day-to-day lives. An IoT device is something as small as a lightbulb that can be controlled using your phone. There are massive objects or devices with smaller IoT devices within them, such as a plane having sensors and gauges monitoring every aspect of the aircraft to ensure it is functioning as expected, keeping the aircraft safe to operate. IoT devices have much potential in helping tourists and making the industry a much more friendly place in the tourism field.

9.1 INTRODUCTION

Kevin Ashton first coined the word in 1999 when he was giving a presentation. The first IoT device was created in 1990 when John Romkey connected a toaster to the internet with the help of TCP/IP control. This toaster could lower the bread to begin toasting and even turn it off, and the bread would come back up all by only using a laptop connected to the internet. People were swept away by this. They found it quite interesting. Not everything was smooth sailing because people were concerned for their security and privacy. One of the top questions asked was how many people can see my toaster online and control it. John said that this was not a significant issue since the toaster is such a simple device, and unless you were concerned about people finding out about how much you were toasting, privacy was not a concern.

On the other hand, security could be an issue since, in theory, anyone can turn on your toaster while you are not home, which would keep it running and might even start a fire. Some people argue that the first IoT device was a Coca-Cola vending machine in Carnegie Melon University. Students embedded sensors inside the machine to let them know whether the Coca-Cola drinks were being cooled enough. Moreover, this marvelous invention then catered to future IoT research. However, the concept of connected devices originated back in 1832 when the first electromagnetic telegraph was invented.

The whole concept of the Internet of Things (IoT) has been exciting to me. The very fact that we can control our door locks, air conditioning temperature, and even your smartwatch alerting you about your rising blood pressure has caught my attention. The importance of IoT includes its influence on industries. It is way beyond just sharing and exchanging data. It can power whole cities, make them smart, and even wholly automate factories, reducing labor costs. In order to understand its influence, one must talk about the industrial Internet of Things (IIoT). It is defined as the automated interaction between intelligent IoT devices and systems that exchange data to either an office or a cloud solution for analytics using devices that use meager amounts of energy, are very easy to install, and are up to industry standards. Numerous industries are starting to use this technology to a greater extent. Cities are making traffic management automatic, which reduces and monitors traffic flow, making it quite efficient and straightforward to cruise the city. Sensors in parking garages can help find a parking space through a parking application, making it very easy and intuitive for new visitors or tourists. IoT just does not mean turning a light on with your phone. The actual depth of IoT is way more profound and filled with potential. However, gathering the data from all these devices and improving them can come a long way in developing the technology. This amount of data allows the consumers, organizations, and even entire smart cities to run cleaner. However, collecting data at this amount does have its problems. The purpose is to focus on the potential this holds in the tourism industry.

Artificial intelligence (AI) and IoT have revolutionized the Tourism industry in the last decade. The shift from the innovation in the booking portal has moved to the pre-arrival choices. Furthermore, analysis has impacted the pre-arrival to the experience of the tourist. Day by day, the growth of IoT is tremendous in many fields such as transport management [1–3], supply chain management [4, 5], tourism, and hospitality [6, 7], agriculture management [8, 9], healthcare [10, 11] [12, 13]. This has provided a better service to tourists and has impacted the efficiency of trains, flights, and hotels [14]. AI has already impacted the business of the Travel and Hospitality industry, and it is expected to exceed $1.2 billion by 2026, with the CAGR estimate of 9.7% by 2026 [15]. The hospitality industry is exceling with the influence of technology, and China is leading the industry in robot hotels and performing efficient operations [16].

9.2 LITERATURE REVIEW

Smart Destination is the future of tourism in the advancing cities with technology-enabled frameworks in all its operations, especially for the cities claiming themselves as the top smart cities. The governments are in the perspective of becoming a Smart Destination (smart city willing to portray itself as a tourism destination). This is possible with the basic infrastructure of the smart city itself. It must find a link between tourism and trending valuable technologies in the city and parallelly implement the strategies to enhance the tourist's value [17]. This will enhance the proper utilization of the smart city resources and boost competitiveness among similar destinations, which translates into the identification, development, and implementation of the proper technological aspect in the appropriate processes to enhance efficiency, comfort, and sustainability in broad terms. It will entail the sustainable development in economic, environmental, and socio-cultural aspects of the tourism industry, tourist, and city itself to boost the sustainable development of processes, city, and quality of life of locals and tourists. This will also enhance the understanding of marketing and production processes; its flaws will be removed, and an efficient system will be in place to appropriately manage crucial segments such as employment and tax revenues and overall satisfaction [18].

Next generation is expected to be more intelligent, and with the influence of innovative technology booming the industry, smartness has got several forms and shapes in each domain. The inception of the smart city has brought smart destinations and led to innovative destination management through the comprehensive analysis of data collected from the overall value chain of the tourism industry. This will allow service providers to get to know the individual's choices, Personalize the product as per the choices, analyze the data from several services, and compare the outcome in real-time [19].

The IoT is continuously gripping every department and section of the industry, such as at the airport, travel sites, hotels, kitchen, train, Flight (stated as the internet of Hotels, Internet of Flight, Internet of Railways, etc.). These examples are already in operation and have started providing real-time benefits to the service providers and the users [20].

The personalized hotel rooms offer smart keys in smartphones, preferred TV channels, lighting, and temperature settings of each guest, with the voice-based interaction devices throughout the stay. IoT provides body-based sensors, revised inventory management, automation, and monitoring of the tasks in the hotel building [21].

Around 80% of the hotel operators in one of the surveys conducted by oracle confirmed that the hospitality industry is adapting the IoT-based voice assistant to control the room's devices and the temperature and lights [22]. The IoT solution, with its varied capabilities, will be used in housekeeping and maintenance activities through motion, leak, or vibration sensors added with RFID, Wi-Fi, and beacons [23].

9.3 EXAMPLES WITHIN TOURISM INDUSTRY

While many industries already use the IoT, the tourism industry, in particular, is very well set up to take advantage of IoT. It can help make doing day to day smoother, like running a hotel or travel company.

9.3.1 MORE CONTROL

The topmost consumer usage of the IoT technology in the tourism industry is personalized services throughout their stay or travel with a single device managing several operations based on individual choices in operating appliances, services, and access [24]. The customer is enabled with the choices of switching on or off, navigating to the device options and choosing appropriate options for his stay accordingly by his phone or tablets. The option is not limited to television and music only. However, it covers the temperature of the bathing water and the intensity of light in the particular lamp as per the individual's choices at a certain time of the day. This connectivity with the internet would regulate the flight seat temperature or air conditioning and enable customers to be more connected.

9.3.2 EFFORTLESS TRAVEL

Travel has been a significant part of the tourism industry, and the IoT has simplified it in several portions by smoothening complex processes. In order to trace a passenger inside the airport becomes difficult, and conveying him the appropriate information at the right time makes it more crucial. For example, sending messages to arrival tourists for their baggage arrive at the baggage belt can be easily communicated directly to passengers' smartphones by tracing them on the sensors and notifying the baggage location. The electronic delivery of the room key for the hotel guests in their smartphone can be activated upon arrival at the reception desk without waiting for any action and letting them check-in automatic in an absolute sense. Sensors might also be utilized to notify restaurants and transportation in several other tourist locations to get similar on arrival check-in or transactions using sensors only.

9.3.3 REPAIRS AND DOWNTIME

The commonly quoted downside of the technology is the runtime errors, breakdowns, and repairs leading to the downtime. Maintenance is inevitable to get uninterrupted services from the installed technologies without any fatal stoppage. The same IoT could support all devices with real-time device information about its remaining life, present state, etc. This is critical for the tourism industry in its railway and flights operations, where a simple error may be hazardous. The IoT is used to identify faults in railway or flight engines and effectively supports the stoppage of any such rundown of the systems. This functionality contributes to the efficiency, maximizing the value and safety of passengers.

9.3.4 ENERGY EFFICIENCY

Efficient systems are primarily based on the energy supply and conservation at all possible ends. The big plants and organizations have 24×7 operations and energy consumption at several ends. A longer idle time is recorded between the active usage of the machine, lights, fans, and many other electronic types of equipment. This generates a lot of energy consumption in terms of unnecessarily heating and cooling the devices. The IoT effectively saves the consumption of energy resources based on equipment availability and its specific temperature requirements. Similarly, the high-consumption lights can be adjusted with the availability of natural light with the help of sensors.

9.3.5 GEOGRAPHIC POSITION-BASED INFORMATION

The tourism companies are utilizing the IoT in its comprehensive frame by providing messages based on the geo-position of the tourist. This provides multi-facet benefits by providing information to tourists and the service providers about the real-time choices, time spent, and honest feedback from tourists, which may be highly productive for analyzing the tourist's behavior and adjusting the services accordingly. The beacon and sensors can be integrated with smartphone capabilities to provide the necessary information, offers packages, attractions, and other essential features to make the travel truly unique and satisfying to each individual.

9.3.6 TOURISM EXPERIENCE

Tourists will simply identify and get information on anything they need when visiting the city using their cell phones. Sensors will be installed around the city, from bus stops to parks and gardens to parking lots, to collect data on travel habits and reactions to various attractions, identify patterns and enhance the visitor experience [25].

9.3.7 AUTOMATION

The IoT automates travel and lodging industry processes. Through sensors in shipments, hotels can better manage supply chains, allowing them to prepare for any potential scenario and minimize guest service interruptions. With the centralized management of these IoT-enabled cameras and sensors from smartphones, hotels, and airlines may quickly establish security mechanisms in offices and buildings [26].

9.3.8 LIVE INFORMATION

IoT apps will aid travelers in connecting with the many devices they bring with them. Real-time information on the flight status will be accessible if you need to change flights or connect flights. You will also receive directions to your airport gate, instructions on how to board an aircraft, and details on which security checks have been completed.

9.4 CHALLENGES AND OPPORTUNITY WITH AI AND IOT IN TOURISM

The challenge is to implement these IoT devices in the tourism sector to help comply with today's health regulations of social distancing and COVID-19 restrictions. Thankfully, IoT technology provides a wide range of solutions for tourist sites and businesses to attract new tourists, increasing client loyalty via the confidence provided and, of course, increasing the value offering to generate additional revenue [27].

Because many people are entering and exiting a country, it can be difficult to screen everyone for any symptoms of COVID-19. Devices with thermal scanners and cameras can be kept in public places to screen individuals for their temperatures. If an abnormally high body temperature is recorded, the authorities will be alerted so that the individual will receive medical attention.

Every hall or store has a maximum capacity since social distancing was implemented. However, it is impossible to count every person who enters and leaves manually. So here, an IoT device can be used at the entrance to count incoming and outgoing tourists, and when exceeded, it can alert the store manager to take the required action.

As discussed earlier, a parking detection app can be used to help tourists navigate around parking lots and garages to find a parking spot faster.

Chatbots and virtual assistants are also helping to improve customer satisfaction. Travelers are being prompted by using AR and VR to journey to exotic places and determine the specifics on their own. The IoT is another technology that is helping to advance this business (IoT). IoT is also assisting the tourism sector in a variety of ways.

For example, smart gadgets may be used to improve the physical condition of a hotel and its rooms while simultaneously lowering energy costs. In the case of flights, IoT sensors may be used to alert flight attendants if a passenger's anxiety levels are rising over a certain threshold. In addition, IoT can provide travelers with improved control and access to data via their smartphones. Allowing for a higher level of customization in hotels and flights is a critical use of IoT in the travel sector. This has been made possible by allowing users to control many appliances or services from a single location using centralized devices such as tablets or smartphones [28]. Customers may turn on or off elements like lighting and heating enabled via the internet from a single place when they are implemented. Additionally, customers may choose a specific temperature and light level, with the gadget unwittingly keeping these settings. This technology may also be used on flights to regulate seat temperature and lighting.

9.5 CONCLUSION

The IoT is still in its early stages of development, but it will revolutionize the tourism industry. Customer service will improve due to IoT, as will revenue and customer loyalty. We will have access to large volumes of data due to IoT, but this data will need to be analyzed and comprehended. Therefore, optimal solutions will require investments in relevant technology. For travel firms to profit from future breakthroughs, they must immediately begin integrating IoT into their systems. IoT is enabling users, travel companies and tourists and will reform all the old processes into a new tech-driven framework. The governmental and private entity collaboration in the IoT-enabled system will turn the possible growth and utilization. This will soon become a genuine need and necessity for them, not just something they expect. Nonetheless, suitable communications and data management infrastructure is required for IoT adoption in tourist areas. Tourist operators will have access to vast amounts of data, but this data must be analyzed and comprehended. It will also be required to have data management skills and information quality control systems. Companies in the travel sector must begin adopting IoT into their systems as soon as feasible to profit from future advances. IoT technology is critical in the tourism business, and both travelers and management are adjusting to this new era rather quickly.

KEYWORDS

- **artificial intelligence**
- **automation**
- **geographic position**
- **Internet of things**
- **sensors**
- **tech-driven framework**
- **technology**
- **tourism**

REFERENCES

1. Madana, A. L., & Shukla, V. K., (2020). Conformity of accident detection using drones and vibration sensor. In: *2020 8th International Conference on Reliability, Infocom Technologies and Optimization (Trends and Future Directions) (ICRITO), Noida, India*, (pp. 192–197). doi: 10.1109/ICRITO48877.2020.9197783.
2. Siraj, A., & Shukla, V. K., (2020). Framework for personalized car parking system using proximity sensor. In: *2020 8th International Conference on Reliability, Infocom Technologies and Optimization (Trends and Future Directions) (ICRITO), Noida, India* (pp. 198–202). doi: 10.1109/ICRITO48877.2020.9197853.
3. Madana, A. L., Kumar, S. V., Sharma, R., & Nanda, I., (2021). IoT-enabled smart boarding pass for passenger tracking through Bluetooth low energy. In: *2021 International Conference on Advance Computing and Innovative Technologies in Engineering (ICACITE)* (pp. 101–106). doi: 10.1109/ICACITE51222.2021.9404602.
4. Wanganoo, L., & Shukla, V. K., (2020). Real-time data monitoring in cold supply chain through NB- IoT. In: *2020 11th International Conference on Computing, Communication and Networking Technologies (ICCCNT), Kharagpur, India* (pp. 1–6). doi: 10.1109/ICCCNT49239.2020.9225360.
5. Wanganoo, L., Shukla, V. K., & Panda, B. P., (2021). NB-IoT powered last-mile delivery framework for cold supply chain. In: Singh, T. P., Tomar, R., Choudhury, T., Perumal, T., & Mahdi, H. F., (eds.), *Data Driven Approach Towards Disruptive Technologies. Studies in Autonomic, Data-driven and Industrial Computing*. Springer, Singapore. https://doi.org/10.1007/978-981-15-9873-9_22.

6. Amit, V., et al., (2021). *J. Phys.: Conf. Ser.* 1714 012037 – repeated adjust this.
7. Wonda, G., et al., (2021). *J. Phys.: Conf. Ser.* 1714 012032.
8. Shukla, V. K., Kohli, A., & Shaikh, F. A., (2020). IoT based growth monitoring on *Moringa oleifera* through capacitive soil moisture sensor. In: *2020 Seventh International Conference on Information Technology Trends (ITT)* (pp. 94–98). doi: 10.1109/ITT51279.2020.9320884.
9. Nanda, I., Sahithi, C., Swath, M., Maloji, S., & Shukla, V. K., (2020). IIoT based smart crop protection and irrigation system. In: *2020 Seventh International Conference on Information Technology Trends (ITT)* (pp. 118–125). doi: 10.1109/ITT51279.2020.9320783.
10. Athota, L., Shukla, V. K., Pandey, N., & Rana, A., (2020). Chatbot for healthcare system using artificial intelligence. In: *2020 8th International Conference on Reliability, Infocom Technologies and Optimization (Trends and Future Directions) (ICRITO)* (pp. 619–622). doi: 10.1109/ICRITO48877.2020.9197833.
11. Trayush, T., Bathla, R., Saini, S., & Shukla, V. K., (2021). IoT in healthcare: Challenges, benefits, applications, and opportunities. In: *2021 International Conference on Advance Computing and Innovative Technologies in Engineering (ICACITE)* (pp. 107–111). doi: 10.1109/ICACITE51222.2021.9404583.
12. Shukla, V. K., & Amit, V., (2019). Model for user customization in wearable virtual reality devices with IoT for low vision. In: *2019 Amity International Conference on Artificial Intelligence (AICAI)*. IEEE.
13. Shukla, V. K., & Amit, V., (2019). Enhancing user navigation experience, object identification and surface depth detection for "low vision" with proposed electronic cane. In: *2019 Advances in Science and Engineering Technology International Conferences (ASET)*. IEEE.
14. Verma, A., & Shukla, V., (2019). Analyzing the influence of IoT in tourism industry. *SSRN Electron. J.* doi: 10.2139/ssrn.3358168.
15. *Travel & Hospitality AI Market Size, Industry Outlook, Market forecast, Demand Analysis, Market Share, Market Report 2021–2026*. https://www.industryarc.com/Report/18662/travel-hospitality-ai-market.html (accessed on 07 September 2023).
16. *AI in Hospitality in China: Are Robots Taking Over the Hotel Industry*, (2019). Daxue Consulting – Market Research China, https://daxueconsulting.com/ai-hospitality-china/ (accessed on 07 September 2023).
17. Ordóñez, M. D., Gómez, A., Ruiz, M., Ortells, J. M., Niemi-Hugaerts, H., Juiz, C., ... & Butler, T. A. (2022). IoT technologies and applications in tourism and travel industries. In *Internet of Things–The Call of the Edge* (pp. 341–360). River Publishers.
18. Ávila, A. D., Lancis, E., García, S., Alcantud, A., García, B., & Muñoz, N. (2015). Smart Destinations Report: building the future.
19. Pencarelli, T. (2020). The digital revolution in the travel and tourism industry. *Information Technology & Tourism*, *22*(3), 455–476.
20. Verma, A., Shukla, V. K., & Sharma, R., (2021). Convergence of IoT in tourism industry: A pragmatic analysis. *J. Phys. Conf. Ser.*, *1714*, 012037. doi: 10.1088/1742-6596/1714/1/012037.
21. Car, T., Pilepić, S. L., & Šimunić, M., (2019). *Internet of Things (IoT) in Tourism and Hospitality: Opportunities and Challenges* (pp. 163–173). doi: 10.20867/tosee.05.42.

22. *Machine Learning and AI in Travel: 5 Essential Industry Use Cases* (2021). IoT For All. https://www.iotforall.com/machine-learning-and-ai-in-travel-5-essential-industry-use-cases (accessed on 07 September 2023).
23. *How IoT Can Benefit the Smart Hotel and Hospitality Industry,* (2021). IoT For All. https://www.leverege.com/blogpost/how-iot-can-benefit-housekeeping-and-maintenance-in-the-hospitality-industry (accessed on 07 September 2023).
24. REVFINE, *How Can the Travel Industry Benefit from the IoT*? [Online] https://www.revfine.com/internet-of-things-travel-industry/ (accessed on 07 September 2023).
25. Nidhi, D. *8 Ways in Which IoT is Shaping the Future of Travel Industry*. Digital Doughnut. https://www.digitaldoughnut.com/articles/2018/january/ways-in-which-iot-is-shaping-the-future-of-travel (accessed on 07 September 2023).
26. Deloitte, *Transportation, Hospitality & Services, Delivering Transportation, Hospitality, and Services Industry Insights to Help Drive Change*. https://www2.deloitte.com/us/en/pages/consumer-business/solutions/travel-hospitality-and-leisure-industry.html (accessed on 07 September 2023).
27. Libelium, *IoT Solution for Tourism*. https://www.libelium.com/iot-solutions/smart-tourism/ (accessed on 07 September 2023).
28. The Internet of Things Report, *Internet of Things*. [Online] https://theinternetofthings.report/resources (accessed on 07 September 2023).

CHAPTER 10

E-Tourism and the COVID-19 Pandemic

SHAKHNOZAKHON FAKHRIDDINOVA

Specialist of the Research and Analytical Activities, Department of the Tourism Innovation and Co-Learning Center, Management Development Institute of Singapore in Tashkent, Tashkent

ABSTRACT

Tourism is one of the fastest-growing industries in the world. At the same time, with the desire of travelers for new brand tourism experiences, climate changes, the rapid global technological advancements, and other dynamics, the tourism industry is constantly transforming to digitalization. The tourism industry faced significant challenges and uncomfortable changes due to the COVID-19 pandemic. Nonetheless, there are still growing digital tourism trends that can help the industry suppliers restart their businesses and experience success until the sector gets into the new normal. Digitalization has played a crucial role in the tourism industry over the last decade. E-tourism has assisted enhance operational efficiency, reducing expenses, and developing services and customer experience. Businesses and travelers benefit from enhanced communication, guest service, and reservation systems. IT has helped the tourism industry replace costly human labor with IT labor. In addition, this can reduce labor costs and avoid customer service problems. Future development of trends in e-tourism is strongly correlated with the development of digitalization. Almost both leisure and business travelers make their arrangements through the internet. It means the industry integrated into digitalization using hardware, software, the internet, innovative technology, and webpage trends. All the above-mentioned digital trends are widely used to improve tourism services and hospitality management.

This chapter presents the current digital tourism trends, focusing on the impact of COVID-19 on digital tourism trends and how they are implemented to recover the industry during the post-COVID period, specified in the case of Uzbekistan and some other countries. The chapter also presents how digital tourism trends shape the industry's future.

10.1 INTRODUCTION

Tourism is a continuously growing and economically beneficial industry in the world. It creates significant opportunities in the labor market, domestic value-added and export revenue, and improves sightseeing's attractiveness. At the same time visit, the country invests and work there. What measures should be taken to build resilience and address the challenges comforting the industry (climate change, pandemic, digitalization, and other extended issues) and ensure the tourism industry benefits from this growth? This chapter outlines current digital tourism trends, the impact of COVID-19 on digital tourism trends, and how they are implemented to recover the industry during the post-COVID-19. This is also specified in the case of Uzbekistan and some other countries, how digital tourism trends shape the future of the industry, artificial intelligence (AI), and its impact on tourism trends. In addition, the chapter will take a deeper look at digital tourism trends [1].

10.1.1 TOURISM AND TECHNOLOGY

IT technology has always played a crucial role in developing tourism. After computer systems became available, they were widely used in the 60s and 70s in hotels and in transportation services to support the function of these sectors. Central reservation systems (CRSs) and global distribution systems (GDSs) such as Amadeus, Galileo, and others started to develop in the 1070s and 1080s. They were first used by airline companies and then by the hotels to maintain reservations and provide price information, travel agencies schedules and many others. The tourism world has massively changed and got massive success with the help of Internet technology. Nowadays, it is impossible to imagine the tourism industry without internet technology because the travel and tourism industry is one of the most critical users of the Internet. At the same time, the Internet has become an essential communication tool for both industry people and travelers.

10.2 THE CURRENT DIGITAL TOURISM TRENDS

Digital tourism has encouraged both local and international competitiveness of the tourism industry, and digital tourism trends are now playing a crucial role in any tourism company in the world. Digital tourism is more oriented toward customer-centric relationship-building management. Digital tourism represents the process of digitalization in the tourism industry, which allows managers to take full advantage of companies' effectiveness and efficiency.

Figure 10.1 displays the fallback of tourist arrivals in the year 2020 compared to the last year's trend in 2019 due to the COVID-19 Pandemic situation, which resulted in a 74% fall in international arrivals worldwide. Digital tourism includes mainly all electronic business purposes such as electronic planning, electronic strategy, electronic procurement, electronic HRM, electronic accounting, electronic finance, electronic marketing, electronic management, electronic commerce, and many others. The rest of tourism enterprises' strategic relationships and business processes are constantly transformed by digital tourism. The present and future development of digital tourism are mainly related to the development of digital technology, including both software and hardware evolution of the technology. According to [6], 60% of leisure and 41% of business-rclated travels are made through the internet. It means that the travel and tourism industry has fully transformed into a digital business. The tourism industry stands where social media, VR technology, insight of data-driven and mobile customers are essential in the digital era.

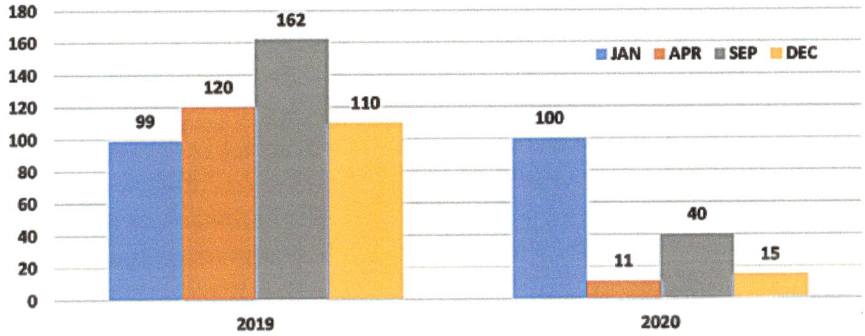

FIGURE 10.1 International tourist arrival 2019–2020 [2].

10.3 HARDWARE TRENDS IN TOURISM

Since the first computers in the 1960s, there have been many changes in digital technology. The first laptops (i.e., Machintosh portable) appeared in 1989 and became lighter, faster, and thinner than ever. The first mobile phones came in 1983 and turned into powerful smartphones. Just one smartphone can merge various technical devices in one. For instance, Microsoft Office, mail, web, electronic book functions, MP3 and video player, powerful digital video camera functions are good examples. Year by year, digital devices are becoming more powerful than previous ones. New devices introducing brand-new technical functions are equipped with a gyro, GPS, Bluetooth, and Wi-Fi connections with the internet and other devices.

10.3.1 INNOVATIVE ACCESSORIES

New hardware trends such as smart bracelets and smartwatches substitute smartphones for convenience in daily life and during travel. First, smartphones and bracelets were purposed only for navigation and exercises, but now they are substituting the typical smartphones with the same functions such as apps and social media. In 2013, Google introduced smart glasses (Google Glass Explorer), which converted into wearable computers with HD cameras, optical touchpad, high-resolution displays, and other powerful functions. Smart glasses have a range of advantages, such as collecting information from the internet, being controlled by voice commands, or recognition by gesture, and tracking activities.

10.3.2 BIONIC AR AND VR

The next hardware development step is Bionic Contact Lenses with added bionics as an augmented reality (AR) with functional infrared lights and electronic tours to create a virtual display. All the hardware, as mentioned earlier, trends have already started and will continue to influence on Tourism industry via the capacity of VR and AR. VR technology creates a virtual world with headphones and takes the user or tourist to the past, future, and new destinations. Smartphone VR (Daydream, Gear), "6 DoF VR" headphones, Tango, Cardboard VR video and 360 cameras, and "You

Visit" apps have almost created virtual tourism as a substitute to real-life tourism. VR devices allow travelers to fly over the skyscrapers in 3D, explore wonders of the earth, surfing in the oceans. VR and AR are also used in cultural heritage tourism to observe heritage sites, sculptures, architecture to receive the necessary information to infer better. VR lets tourists sit on a seat to explore the world, while AR enhances travelers experience on the tourist site. These digital hardware trends link heritage and history to obtain historical events on a heritage spot. For example, AR allows tourists back to any historical period while traveling around the site.

In contrast, VR is used to take the traveler to the historical experience of the touristic destination. Tourists can use a smartphone application or a tablet to experience the place's cultural heritage, while "HTC Vive VR-Googles" allow gaining the travel experience from home. The hardware, as mentioned earlier, digital trends show that VR and AR will become necessary applications to improve the travel experience of tourist destinations. These applications are pretty expensive to produce for the technology; however, VR headphones and AR wearables are getting more widespread and affordable for the tourism industry. Qantas airline is a good example, who was the first among airlines to introduce VR headphones in 2016. The "Natura Reality" was launched by the Nature Conservation Associations SNF for a climate-smart 360-nature experience in reality by utilizing Ski-Googles in the woods.

10.3.3 THE LATEST DIGITAL TOURISM TREND

"Oculus Rift VR" mixes smartphone technology with motion sensors and completes image design on the retina as an old projectors function, introducing the prototype of the virtual world devices. In the future, this technology will allow tourists to travel the world by sitting at their homes and checking various 3D holiday events, transporting themselves to any five-star hotel. In addition, the near future will present haptic technologies for the sense of touch, which allows travelers, feel the texture and the surfaces of the tangible objects. For instance, they can feel the cold of the snow or the warmth of the sun on their hands. Moreover, the haptic technology will allow the hotel guests to walk around the desired hotel to see the rooms, to see how they cook the food, look through the hotel facilities before booking a room in a particular hotel. These powerful technologies are customer-oriented and can build confidence between tourists and tourism suppliers.

10.3.4 3D PRINTERS

Another new hardware tourism trend is 3D printers. Tourism marketing people have already started to use it to produce personalized souvenirs for the visitors. At the same time, action cameras have become well-known among tourists for recording their travel experiences. However, an Eye Tap Technology [1] might need supplementary action cameras very soon because the Eye Tap technology is worn as lenses but works like a camera [3]. It can record the whole scene, available to the human eye.

10.3.5 360 CAMERAS

Trending has double 180° lenses, which merges the pictures into a 360. Before, people took many pictures with their smartphones or cameras and then used software applications to merge images into 180 and 360 images. Nevertheless, 360 cameras allow tourism suppliers and travelers to immediately take an image of surround or film a 360 video, and with a few clicks, download them to YouTube or other social media channels. The current technology gives an excellent opportunity to promote hotel facilities. For instance, rooms, hotel restaurants, gym, banquet, and conference halls, gardens, and the hotel environment allow visitors to turn around 360° in the image. These cameras are good at creating VR presentations of tourism facilities, but these devices have not yet been broadly used by tourists so far.

10.3.6 4D–7D TECHNOLOGIES

A highly experience-oriented technological solution, widely adapted in amusement parks and tourist centers, is 4D–7D movie technology. These technologies allow visitors to feel real experience by wearing VR glasses, and they can go to travel. The main features of these technologies are the movement of seats, special effects, such as wind, snow, bubble, rain, and many other effects. A good example can be the visitor Center in Italy, which has already adopted a 4D complete package with seat movement, water spray, and wind to let their visitors feel the experience of the adventure at the national park.

10.3.7 ROBOTICS

The ultimate hardware tourism trend evolution is robotization, which will play a crucial role in the tourism and hotel industry. It is predicted that the tourism and travel industry will be influenced by intelligent android robots such as EveR. Hitachi developed the EMIEW3 robot, and now it is under test at the Haneda airport, Tokyo. It has a wide range of functions such as helping passengers at the airport, giving information. It can also answer English and Japanese and guide passengers if they lose or cannot find any airport departments.

Surprisingly, robots might replace front office agents, hoteliers, interpreters in the tourism industry. The intelligence, behavior, and appearance of robots have become more human-like. However, it can take up many people's jobs in hospitality; the hotel business has already been run successfully by computer technologies. A good example can be ticket recommendations, automatic flight check-in, google searches, self-reservations, self-check-in, which are non-robotic forms of automated services. Henn-na Hotel in Japan has already functioned robotized hotels where front office employees have unique service faces [4]. Some automating hotel services such as YOTEL [5] with its self-reservations and self-service check-in. Property management system (PMS) automatized by robot helpers, which allows quick access and process of functions such as room management, reservation, setting hotel room rates.

10.4 SOFTWARE TRENDS IN TOURISM

The software trends have been developing with hardware and computer capacity evolution. Windows, Mac OS, operation systems have immensely improved since they appeared first and very soon, AI will be assimilated to any software applications. The evolution of the World Wide Web or the internet increased the demand for various software applications such as web browsers and many others. Netscape launched its first web browser in 1994, and then other web browsers were soon released. They were the following: Microsoft's Internet Explorer, Yahoo! Search, Opera, in 1995, and Mozilla Firefox, Apple's Safari, Google Chrome, Microsoft appeared between 2002 and 2015. According to Gelter, 2017, 3 billion people, is 47% of the world population, using the internet in 2016 [6]. All users use standard methods to access the internet by using the telephone circuit, coaxial cables, copper wires, fiber optics, and cell phones technologies such as 3G and 4G, Wi-Fi, and Hotspots.

All the above mentioned are broadly used in the hospitality industry, and now it is impossible to imagine any hotel without access to the internet. It became one of the main requirements of the hotel guests, travelers, and the modern world to complain directly if there is no access to the internet. 4G software trends enable cell phone internet access, gaming services, IP telephone, high-definition mobile (HDM) TV, video conferencing, and 3D TV, allowing for 100 Mbps and 1 gigabit a second. Huawei introduced CloudRAN [7], a cloud-based solution in 2016, which will support 3G, 4G, 5G, and the connection between things and people. Moreover, 3G, 4G smartphones, personal digital assistants (PDAs) with global position systems (GPS) allow visitors to retrieve necessary information anytime and anywhere. While mobile routers became well-known among transport suppliers, they widely use Wi-Fi in taxis, busses, airplanes to attract more passengers.

10.5 DIGITAL TOURISM DURING COVID-19 PANDEMIC

Having experienced an all-time high in the past decade, with 2018 recording the highest international tourist arrival, according to the World Tourism Organization (UNWTO), the sector has been undeniably faced with big challenges and uncomfortable changes in the face of the COVID-19 pandemic [8]. Nevertheless, there are still rising digital tourism trends that could help industry suppliers to revamp their businesses and experience success as they get into the new normal. Many international hotel chains survived the pandemic; digital tourism trends are engaged to mitigate the pandemic impact on the hospitality industry. An increasing number of digital technologies was the only way to reduce health and safety risks within the industry.

10.5.1 DIGITALIZED HOSPITALITY

During the Pandemic situation around the world, many hotels used live promotions in the live-streaming platforms to increase the number of potential customers and to support them by giving them detailed, relevant information. Not only tourism but many other businesses also survived with social media's help because it was the only convenient and safe way to sell products and reach customers. For instance, famous Chinese video platforms like Kuaishou and Douyin gained more than 4,000 new subscribers in two weeks [9].

International hotel chains and resorts turned to live streaming to continue their work. They organized different online events, exhibitions, and created some online communities to discuss and offer their service and attract more customers. During the pandemic, the world rapidly adapted to new VR and became active participants in e-commerce, e-learning, online entertainment, virtual conferencing, and remote working. However, the hotel industry, restaurants, and travel companies were losing their businesses. After the pandemic shock, the tourism and travel industry also started to adapt to the new reality by conducting international events and conferences to use 5G networks and experience in organizing such significant events. Online conferences gave prospects to business hotel industries more networking and sponsorship opportunities. Numerous international organizations chose to conduct events online. Remote working and online conferences became more common for hoteliers and their guests.

Figure 10.2 depicts the rise in several industries the growth of online digitalized content delivered by B2B in the COVID-19 scenario in Unites states [10]. This shows the growth trends with a difference of a month only, indicating that the usage and adaptation rates were high.

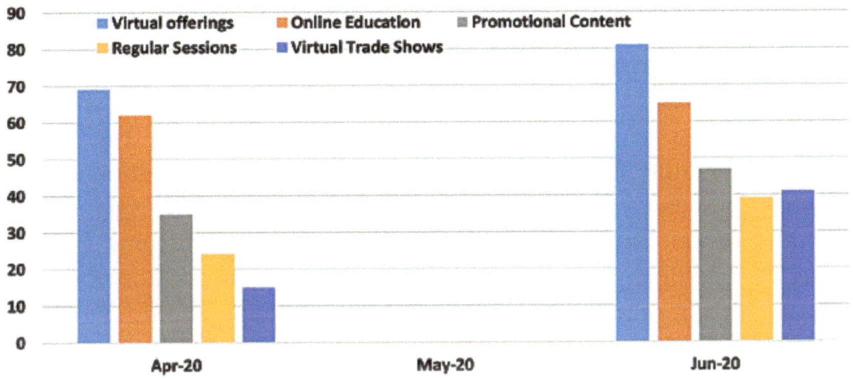

FIGURE 10.2 Digital content during the pandemic – B2B exhibitors [10].

10.5.2 GROWTH IN DIGITAL DEMAND

The pandemic made many businesses, including tourism, turn to digital applications to save their businesses. The COVID-19 pandemic increased demand for solutions of 5G technology. 5G networks have a faster download speed than 4G technology. It provides an excellent opportunity to remote work from everywhere. 5G networks also allow intelligent connectivity and use AI, the internet of things (IoT), cloud and other online platforms. 5G networks handle connections for new apps and devices and support the growing number of new digital apps worldwide. The combination of AI and 5G enables instant check-in and check-out and making payment via facial recognition, a tremendous improvement in tourism service and safety.

10.6 TRAVEL RESUMES IN PANDEMIC

Digital tourism trends allowed developing travel and tourism industry despite the pandemic situation within the world by using a combination of 5G and AI. Recognition technologies with fingerprint or retina scanning are used to unlock guest rooms. This is an excellent option to eliminate physical or key cards. It is not nor surprisingly could the demand to use this digital trend in hospitality during the pandemic increase. The new generation of this technology is facial biometrics, which can automatically be used for check-in and checkout from the hotel and make payments. It prevents social distancing.

During the pandemic, the hospitality industry used robots and AI digital trends to enhance hotel service quality. AI was very beneficial in hotels and resorts. For instance, the quick check-in and checkout processes improved access to the hotel guests experience by using soft wares like facial recognition. Hardware companies of facial recognition created non-contact body temperature measurements to prevent virus spread and meet the market demand. This digital trend was used in some hotels in Guangzhou, China, to save time, reduce overcrowding of hoteliers and visitors. In addition, Guangzhou hotels used robots for the safety reasons of guests and employees during the pandemic. Robots replaced food and beverage (F&B) department employees by doing dining services, working in the restaurants, providing guests with face masks and hand sanitizers, delivering laundry and housekeeping items into the guest rooms. Moreover, they replaced the front office for safety reasons and prevented the virus from spreading within the hotel.

10.6.1 RESTAURANT DURING PANDEMIC

Restaurant businesses and hotel restaurants started food delivery services to survive by using food delivery apps, social media, and online promotion. Across the globe, state bodies try to control the pandemic by using AI for face recognition scanning body temperature, and tracking smartphone data to identify infected people and avoid pandemics. To strengthen safety measures against COVID-19 and to guarantee safe travel, tourism suppliers use facial recognition technologies to measure guests' temperature. The health statuses of the guests are required from the guests in terms of safety and security regulations.

10.7 TOURISM IN PANDEMIC SITUATION: A CASE OF UZBEKISTAN

The pandemic outbreak affected the travel and tourism industry worldwide, including tourism in Uzbekistan. The president of the country – Shavkat Mirziyoyev, signed a presidential decree on urgent measures to support the travel and tourism industry to decrease the undesirable effects of the COVID-19 pandemic. The first measurement stage was to improve domestic tourism among local tourists. In the second stage, receive CIS member countries' citizens. In the third stage, citizens of the third countries. At the framework of the "Uzbekistan, safe travel guaranteed," colossal work has been done including, technology advancements in the industry [11]. The tourism sector in Uzbekistan has already implemented a technological framework; even this was started from pre-COVID-19 time. For example, online bookings like Booking.com and Expedia were popular within the country hotels. Besides this, many hotels used PMS systems such as Fidelio, Opera for hotel management.

However, pandemics made tourism suppliers use digital trends daily, and the usage of e-tourism hit the peak. Almost all tourism-related events were conducted virtually, and the most popular live streaming platform was Zoom in Uzbekistan. Gastronomic tourism was highly promoted in Uzbekistan through online platforms and virtual conferences. For instance, many gastronomic workshops were held in the live-stream mood. Local and international experts, influencers, travel bloggers, and food bloggers were attracted to promote tourism types in Uzbekistan, including gastronomic tourism, via social media. As mentioned above, the world experience in the catering industry is widespread in Uzbekistan. F&B outlets employed food delivery apps and messenger bots as their number one helper during the pandemic. Uzbekistan and worldwide, all businesses survived due to hardware and software digital trends. It is not easy to imagine what would happen if it happened two decades ago. Nevertheless, an enhancement in technology saves businesses, reduces the risk of infection among people, replaces employees at work, and allows people to work remotely during tough times.

10.8 DIGITAL TOURISM TRENDS SHAPING THE FUTURE OF THE INDUSTRY

Even though many people prefer face to face communication, tangible travel experience, authentic atmosphere, the travel and tourism industry will never return to the pre-COVID-19 normal. Travelers and hotel guests will expect higher health safety standards. They will show high interest in contactless tourism services, high demand on the IoT, and high-speed access to the internet. Travelers and guests will be divided into traditional communication and online service lovers. There will always be a demand for e-service within the tourism sector in the future because travelers and hotel guests have already experienced the convenience of digital technology. The hospitality sector will use a diversity of new technologies and business models to improve safety, social distancing, and a comfortable, personalized experience for their guests. Some hotels will change and modify their business models, enhance their service quality and set up 5G network with cloud applications. In the future, 5G smart hotels will open; these lodging facilities will provide AI recognition technology to identify their guests, service robots, e-keycards, while 5G networks will advance the capacity of the online conferences in the hotel.

The catering industry will also improve its delivery service to provide individual visitors and purchase customers. New digital technologies will be launched to enhance safety and hygiene standards in the lodging and catering sectors. Contactless service will become the new standard in the hospitality industry. 5G powered robots, kiosks, self-service apps arranged to run the hospitality in future. The COVID-19 pandemic forced the travel and tourism industry to step forward quickly. The travel and tourism industry will mainly focus on hygiene and safety; tour agents will be more involved in digital marketing and social media. Travel sector suppliers will continue to work remotely by using digital technology. Prepare future tourism employees for awareness of the technology changes in tourism near future. Educating tourism people about computer and digital marketing and the continuous development of e-tourism will make the industry always be in the digital trend. Digital technology will become an undivided part of the tourism sector. Rapid technology enhancement will make the industry always be a step forward among other businesses because technology will be an essential part of the travel industry in future.

10.9 CONCLUSION

The Tourism and hospitality industry is the baseline for several industries and their movements between different locations. The pandemic has resulted in the lockdowns of the countries and made the whole economy break down. The governments took the necessary measures to stop the pandemic, conveying to the population and business that the regulations and their strict implementation became possible with technology and its features, which proved fruitful to everyone. The digital trends and technologies were growing and getting implemented worldwide at their speed, which has spread full throttle in the pandemic situation for communication, verification, and automation of policies, regulations, processes, and safety.

KEYWORDS

- **tourism**
- **digital tourism trends**
- **digital trends**
- **digitalization**
- **digital tourism**

REFERENCES

1. *OECD Tourism Trends and Policies 2020* | en | OECD. https://www.oecd.org/cfe/tourism/oecd-tourism-trends-and-policies-20767773.htm (accessed on 07 September 2023).
2. UNCTAD, "Prosperity for all," COVID-19 and tourism – An update, "Assessing the economic consequences" [Online accessed on 21 October 2023], https://unctad.org/publication/covid-19-and-tourism-update.
3. EyeTap Personal Imaging Lab. https://mannlab.com/eyetap (accessed on 07 September 2023).
4. Brand Concept | Henn-na Hotel Huis Ten Bosch Official Site. https://www.h-n-h.jp/en/ (accessed on 07 September 2023).
5. *Top BTC Trading Platforms 2021.* https://cryptoine.com/top-btc-trading-platforms-2021/ (accessed on 07 September 2023).
6. Martín-Suárez, S., & Encinas, J. M. (2021). The future belongs to those who prepare for it today. *Cell Stem Cell*, *28*(5), 783–785.
7. *CloudRAN: Reconstructing the Radio Network with Cloud.* Huawei Press Center. Huawei. https://www.huawei.com/en/news/2016/4/CloudRAN (accessed on 07 September 2023).
8. *International Tourist Numbers Could Fall 60–80% in 2020, UNWTO Reports* | UNWTO. https://www.unwto.org/news/covid-19-international-tourist-numbers-could-fall-60-80-in-2020 (accessed on 07 September 2023).
9. Fannin, R. (2020). Coronavirus outbreak spurs new innovations in High Tech. East West Bank. URL: https://www.eastwestbank.com/ReachFurther/en/News/Article/Coronavirus-Outbreak-Spurs-New-Innovations-in-High-Tech (accessed on 07 September 2023).
10. *Digital B2B Exhibition Content During COVID-19 US 2020.* Statista. https://www.statista.com/statistics/1199487/digital-content-of-b2b-exhibitions-during-the-coronavirus-us/ (accessed on 07 September 2023).
11. Erkinbek Turgunov, (2020). Weekly-e-bulletin-06.07.2020-12.07.2020-No-267. https://eurasian-research.org/wp-content/uploads/2020/10/Weekly-e-bulletin-06.07.2020-12.07.2020-No-267.pdf (accessed on 07 September 2023).

CHAPTER 11

Artificial Intelligence for Smart Medical Tourism

ANKITA SHARMA

Student, Global Business School, Dubai, UAE

ABSTRACT

Artificial intelligence (AI) and robotics technology are disrupting every industry, transforming traditional business practices, and delivering services. Technologies like machine learning analyze big data to learn human behavior and patterns to make informed decisions to improve customer service experience. AI technology has impacted the hospitality industry through guest data analysis and robotics for providing various services quickly and efficiently to optimize customer satisfaction. The Healthcare industry, too, has been transformed by using AI technology to study patients' medical data and robotics in administration and surgical workflows for better direct patient care. This chapter explores the various usages and applications of AI and robotics in hospitality and healthcare. Furthermore, discuss the AI impact on the medical tourism industry.

11.1 HOSPITALITY INDUSTRY AND ARTIFICIAL INTELLIGENCE (AI)

One of the most significant breakthroughs of technological advancement in AI is becoming more reliable, enhancing its reputation for carrying out various tasks even in the business world [1–4]. Computers and machines

Artificial Intelligence for Smart Technology in the Hospitality and Tourism Industry.
Vinod Kumar Shukla, Amit Verma, and Jean Paolo G. Lacap (Eds.)

advanced and intelligent behavior which collects, stores, and analyzes data, thereby facilitating making well-informed decisions on their own by copying and learning from human reactions and interactions. AI-powered services have brought disruption to many industries and businesses. Everything is available in the palm of our hand, be it in the form of a smartphone, smartwatch, or virtual assistant like telling Siri on iPhone or Alexa at home to call someone or to change the music volume. AI tools can recognize and translate different speeches and languages, identify objects and people and learn from their behavioral patterns, all of which help make better decisions and provide solutions and answers, thereby helping to improve customer experiences.

During the last few years, when the coronavirus pandemic has impacted all industries, the hospitality industry was one of the hardest-hit industries due to restrictions on travel across the world [5, 6]. With the use of robotics and AI, which provide tremendous growth opportunity for the industry wherein people now want lesser human interaction, AI and robotics is the best alternative and solution. Automation is playing important role in industries [7–11]. Companies in the hospitality industry are increasingly using AI to carry out important tasks related to personalized customer service, which is of utmost importance to hotels and resorts. The hospitality industry can create a more seamless customer experience at the hotel, which can help drive more growth and revenue using AI. AI helps in analyzing important customer data to interpret patterns. This can help hospitality companies a competitive advantage over others in the industry. By improving customer service and experience, hospitality-based organizations can attract new customers while retaining the existing guests and customers.

AI and AI-powered robotics have the power to transform the business processes, services offered and facilities provided in the hospitality industry. Automation of existing processes, use of intelligent chatbots and voice-enabled services are some examples where AI has changed the landscape of the traditional hospitality industry. And implementing AI can help in enhancing customer experience and business reputation. As more and more people become familiar with and use more digital devices with AI at home, their need and demand for wanting AI smart experience when visiting hotels increases providing opportunities for growth for hotel owners. AI helps in extracting valuable touch points by analyzing consumer data. And this data can help hoteliers minimize time and effort, reduce operating costs for consumer support functions, maximize workforce efficiency, and improve customer service by handling guest requests, booking reservations, or ordering food and amenities easily and accurately [12].

The hospitality industry focuses on providing the best customer experience, which is the critical driver of growth and guest loyalty. AI can provide responsive, fast and customer-centric hospitality as it can streamline processes and provide valuable insights, which helps to improve guest service and optimize experiences. Automation and AI are revolutionizing the hospitality industry. AI helps hospitality companies with segment automation to target customers based on guests' profile data [12].

Everyday hospitality companies generate data such as personal guest information, food and amenities orders, billing and payment detail. AI algorithms analyzes can analyze this data to help hotels save money by improving operational efficiency and service, be it through automated telephone answering systems or marketing messages through mobile apps by predicting surges in utility demands. AI-powered services can benefit hospitality companies through client behavior modeling and predictive analysis, which studies the traveler's hotel stay data right from before stay to during stay to after stay and generate recommendations that improve customer experience and improve hospitality business. To optimize the use of AI, hospitality companies must balance human interaction with clients. Along with AI, empowered robotics technology interaction with guests rather than implementing multiple AI solutions at once while maintaining the confidentiality and integrity of customer data to ensure data privacy is of utmost importance for companies to take care of [13].

Using AI, hotel owners can reduce and eliminate the element of human errors, which not only helps deliver superior service to guests but also helps save money by preventing unwanted costs. One of the keys to delivering a superior customer experience in the hospitality industry is responding to guest queries faster, which is possible through the use of AI, which can effectively learn, predict, and adapt to customer requests faster. Thereby significantly improving customer interactions experience by fully understanding the customer journey and delivering what the guests are looking for, resulting in improved customer service, which is the key to success for any hospitality company. Understanding and predicting what guests want and providing customized services and product offerings accordingly to fulfill their requirements properly at a faster pace with the help of AI helps hotel management improve overall operations and increase sales by making correct data-driven decisions [14].

11.2 HEALTHCARE AND ARTIFICIAL INTELLIGENCE (AI)

AI and IoT is transforming industries and changing business processes in every field worldwide [15] [16]. The healthcare industry is one such area where it can radically alter the way traditional healthcare services are delivered. Analyzing data about patient medical history/clinic visits, medicines prescriptions, lab test results and reports, and procedures that contain valuable and sensitive health information can transform patient care and diagnosis [17].

Healthcare is one of the most important sectors of society, and the use and application of AI technology for analyzing medical data can be a matter of saving someone's life and delivering quality. Effective patient care is the most important aspect for healthcare organizations. AI technologies can do routine administrative work, which can help nurses, doctors, and other healthcare workers. AI usage in the healthcare industry can provide better preventive recommendations and improve quality of life while delivering more accurate diagnoses and, in turn, more accurate treatment plans resulting in better overall patient care. AI can also help prevent and combat the spread of infectious diseases like coronavirus by analyzing government and healthcare data to track and predict pandemics and thus act as a useful tool for global public health care [17].

Using AI and machine learning algorithms to study medical data helps in patient medicinal diagnosis. Machine learning along with the AI helping in producing relevant outcome for the process [18–20]. AI in healthcare helps doctors and medical service providers to provide more accurate diagnoses and recommend proper treatment plans. AI can help make accurate predictions by analyzing medical data, which helps healthcare providers be more proactive and provide improved quality preventive recommendations for better patient care. AI in healthcare can thus support better clinical decision-making by analyzing medical data. AI can also help improve surgical performance by eliminating variations in outcomes of complex surgeries. AI can improve the efficiency of surgeons by providing them 3D magnification of articulation and helping in precise cutting and stitching. So not only AI-powered robots can help understand surgical workflows or execute complex surgeries. They can also be useful in rehabilitation, long-term illness, physical therapy, adult care, and end-of-life care [21].

AI can also help to improve operational efficiencies in healthcare organizations by optimizing administrative and clinical workflows. AI-powered tools can enhance the productivity of healthcare workers, reduce human errors, and improve clinical and administrative efficiency. This helps reduce operational costs for healthcare and hospital practices and services and improves the service quality experience of patient care delivered. Repetitive tasks performed by healthcare support staff and practitioners can be automated with the use of AI technology-powered robo. These workflows can be automated by using AI, such as machine learning. Be it admissions, billing, administrative documentation, transcripts creation in AI. AI usage in claim processing, clinical documentation, medical records, and revenue cycle management can help reduce data verification time and errors, thereby saving money and resources [22].

AI technology in healthcare can make healthcare services affordable by reducing operations costs, making clinical trials more efficient, improving success rate with accurate diagnosis and effective medicines recommendations, and lifestyle change recommendations through innovative medical applications and devices. AI technology helps people stay healthy as healthcare applications encourage healthy lifestyles and behavior with heart rate monitoring and tracking cardiac health and smart devices having features like fall detection. Emergency SOS helps people be aware of their health status continuously to detect illness early. AI technology-powered smart health devices and health applications can help people stay fit and healthy and, at the same time, provide data every day to healthcare professionals who can remotely track and understands the needs and health behavior of patients to provide timely guidance and feedback to assist them to stay healthy. AI technology can detect diseases like cancer more accurately at early stages by enabling the translation of mammograms faster and more accurately, reducing the need for biopsies. Not only cancer smart consumer wearables combined with AI technology can detect early-stage heart disease, which helps medical professionals like doctors to detect life-threatening diseases early at a more preventable and treatable stage [23].

AI technology can also help provide training to nurses and interns by simulating natural environment situations from a large database of case studies and scenarios, helping trainees to respond to questions and make challenging decisions on the spot, which can be continuously evaluated, remodeled, and adjusted according to trainee learning needs. AI technology-empowered digital devices can help healthcare workers to undergo or continue their training from anywhere anytime, facilitating continuous upgrading of knowledge and skills of healthcare workers, which results in providing consistent service experience for all staff members. Thereby improving the quality of patient care delivery and enhancing customer satisfaction, which is of utmost importance for healthcare organizations.

AI technology usage in the healthcare industry can be in many forms, like machine learning, natural language processing (NLP), etc. Machine learning can be used to predict what treatment procedures can be successful for patients based on their medical data, and a treatment framework can be developed based on the supervised learning and analysis of training data for which results are known. Disease diagnosis and medical treatment are at the center of AI usage in healthcare. Integrating it with clinical workflows and electronic health record systems is key for accurate suggestions. Machine learning is replacing rule-based systems to use and analyze data using medical algorithms because as several rules grow, it begins to conflict, and rule-based systems lose effectiveness [24].

AI can simplify the daily routine of doctors and healthcare administrators by performing tasks in less time and with more accuracy, which effectively helps in patient care service quality. AI can be used in healthcare to discover new links between genetic codes through machines that can learn and predict on their own, thereby reinventing healthcare services and maximizing healthcare organizations' efficiency. Rule-based expert system based on variations of if-then rules by using AI technology is widely used in supporting clinical healthcare decisions, which many electronic health records systems make available. Another AI technology used in healthcare is NLP, which includes speech recognition, text analysis, and translation to understand and make sense of clinical documentation and classification. NLP systems can use AI technology to analyze patients' clinical notes, which are unstructured and give meaningful insights to improve quality, methods, and results for better effective patient care. So be it assisting doctors and surgeons in aspects of patient care or helping other healthcare workers in simplifying administrative work processes, AI technology can help to improve upon existing solutions and overcome medical challenges faster [22].

11.3 ARTIFICIAL INTELLIGENCE (AI) AND MEDICAL TOURISM

One of the fastest-growing industries in modern times is medical tourism. People travel to different places or countries for medical reasons and seek medical assistance and medical care services that are better and less expensive than their place or country. Since the main purpose of such people traveling to other places or countries is for medical reasons, they are also referred to as medical tourists. One of the main purposes of medical tourists is to avail better healthcare and medical facilities and services at affordable costs. One of the main reasons for promoting medical tourism is to provide medical services for patients worldwide at low costs while offering specialized services, surgical treatments, and other specific treatments in coordination with the hospitality and tourism industry. Medical tourists also bring value to the tourism business, so many cities and countries promote medical tourism given the income and revenue generated [25].

Technology advancements have reduced the distance between healthcare providers and patients around the globe who can access healthcare and medical professionals irrespective of geographic location, thereby increasing the medical tourism concept. These facilities generate data about medical tourists, be it planning for traveling by air, booking hotel accommodations, arranging hospital admissions, booking physicians appointment, etc. The medical tourism industry's popularity is growing worldwide, and computer science techniques are needed to utilize a huge amount of data produced online. Medical and tourism data need to be analyzed using AI technology by applying machine learning algorithms to identify trends and patterns and medical tourists' preferences. AI can also help analyze medical tourists' preferred treatment conditions like dentistry, cosmetic surgery, and weight loss surgery. Analyzing such data helps healthcare organizations improve their processes and offer services that will help them attract more medical tourists.

One of the primary concerns for any medical tourist seeking medical or surgical treatment in another place or country is the unknown quality of care and service they will receive from healthcare organizations and medical professionals there. Usage of AI technology by healthcare organizations can help provide quality service and patient care that medical tourists seek that is reliable and value-for-money healthcare services from a modern reputed healthcare organization. So, when medical tourists analyze the options based on prices of different treatment plans, credentials, and degrees of medical professionals, healthcare organization ratings and patient reviews, and description of services offered. AI technology integration can match medical tourist requirements of best services and help recommend global patients with suitable healthcare facilities of that organization.

AI and robotics are transforming every industry, and the healthcare industry is no different as it is changing the global healthcare ecosystem, which in turn is impacting and changing the global medical tourism industry landscape. AI influences patients' choices when deciding which healthcare organization to select to avail of healthcare facilities and services, be it for consulting, testing or treatment. The population is growing globally, and so are the medical requirements. Leading healthcare organizations are trying to offer the best treatments to patients globally, irrespective of geographical borders. They, therefore, are trying to make the best use of introducing and integrating AI technology and robotics into their healthcare support system service, offering to emerge as a preferred medical tourism destination.

Machine learning, NLP, and AI technology can empower healthcare organizations to handle and treat patients with better care quality, thereby enhancing the patient experience and satisfaction levels, which is the key factor for medical tourists. AI technology and robotics integration in global healthcare service offerings are already impacting the medical tourism industry. Professional efficiency, treatment quality, saving time and costs, etc., greatly influence medical tourists. Implementing and integrating robotics with AI technology will help take fast actions for better diagnosis. Smarter précised treatment solutions, which in turn will help the medical tourism sector. AI technology helps doctors or surgeons make decisions, diagnosing diseases while handling huge amounts of medical data. Medical tourists from different countries have chunks of unstructured medical records. AI technology can analyze and help medical professionals save time and make it convenient for them to understand better and treat patients abroad [26].

Medical tourism aims to maintain and improve one's health and well-being by traveling abroad to avail of the best modern facilities and healthcare services. Hospitals, healthcare organizations and travel agencies use artificial technology to analyze data and target patients across the globe with customized ads. Thus, people who look for medical procedures can be targeted with advertisements with the help of AI technology, which will increase the number of medical tourist visits for healthcare organizations. Countries offering world-class treatments from the best medical professionals and specialist experts, affordable healthcare services, and pleasurable stays are preferred destinations for medical tourists. Medical tourism can help earn foreign exchange and create employment opportunities for the medical tourism destination. Countries can provide affordable and reliable healthcare solutions and treatments for medical tourists, which is possible through the proper use and integration of AI technology and robotics. AI technology can also help patients and companions with transportation, accommodation, food, etc. During the pandemic, robotics and artificial technology helped in remote treatment. They made complex surgeries look easier and less invasive, which results in less pain and a faster recovery period, which is beneficial for patients seeking treatment as it helps them save costs. AI and robotics have helped and played an essential role in reviving medical tourism. It has helped patients reduce the need to go to doctors, avoiding long, crowded queues and human contact. This helps prevent and stop infection, which is important, especially during COVID-19 [27].

11.4 ARTIFICIAL INTELLIGENCE (AI) APPLICATIONS IN HOSPITALITY

Hospitality Industry also got benefits from the AI applications and there are number of different applications now such as chatbot, translation, dynamic pricing, operation management, understanding, and analyzing customer behavior. This data helps the hospitality industry to offer more customized services for the enhanced customer experience. The few examples of AI supported application utilized by hospitality industry are discussed in subsections.

11.4.1 CHATBOT

AI technology-powered chatbots are automated software that can automatically respond and reply to guest queries to improve customer service [28, 29]. Chatbots help hospitality companies to provide optimal front-facing customer service to clients while freeing up other human resources for important tasks. Chatbots can answer guests' queries, handle housekeeping requests like room cleaning or getting things like extra pillows or blankets delivered, can handle concierge services, online chat services, manage reservations, etc. So, these queries get solved easily and, more importantly, quickly, providing guests with timely, consistent service while freeing up human resources for complex tasks, which is key to customer experience satisfaction in the hospitality industry [30].

11.4.2 DATA ANALYSIS

Real-time data analysis can be done easily with the use of AI as market surveys, customer feedback data and performance reports can help improve different business aspects. Machine learning makes it easier for hospitality companies to collect, store, and analyze data which can be quickly interpreted, and conclusions can be drawn to make data-based decisions to improve performance. Understanding current guests, as well as potential target guests, helps hospitality companies to attract and retain customers by providing services as per their needs. Hospitality companies can get a holistic view of expenses, profits, revenue, etc., thereby maintaining business cash flow at the top of the market. Data analysis helps in personalized marketing to improve the customer experience so hospitality companies can understand guests' preferences [31]. Data analysis also helps to integrate loyalty programs and improve customer service, helping hospitality companies retain guests. AI can enhance business reputation as it can analyze and amplify positive social media posts and reviews while helping to rectify negative feedback early [32–34].

11.4.3 SERVICE IN VARIETY OF LANGUAGES

Hospitality companies deal with various guests and travelers from different countries speaking multiple languages, which is not known by everyone in staff employees serving them. AI technology-powered chatbots can speak various languages as programmed, or having a live translation app helps to adopt and understand audience language, which will help hospitality companies to serve them well by clearly understanding their needs. AI technology to serve a multilingual population helps hospitality companies attract various guests without having the headache of hiring employees who can speak multiple languages. Breaking communication barriers between business people and clients is the key to providing precise customer service and having a global audience [30, 34–36].

11.4.4 VOICE ACTIVATED SERVICES

AI-powered in-room technology provides guests with a smart home experience, providing smooth guest services and enhancing guest stay experience. Having voice-enabled services like dimming the lights, switching on fans or ac, turning on the music, setting the alarm or wake-up call, etc., provides guests with a seamless experience leading to customer satisfaction as their needs are met and fulfilled instantly. Hospitality companies can capitalize on the growing demand for digital assistant devices in hotels to elevate guests' stay experience with voice-enabled services [34, 37].

11.4.5 HYPER DYNAMIC PRICING

Hospitality companies can benefit from AI by using booking engines to provide hyper-dynamic pricing wherein prices are automatically updated or adjusted based on guest data, potential guests' information on their social media, market data, and competitors' pricing. By understanding current demand trends and studying potential guests' data, AI technology helps offer a room. It offers that will attract and tempt them into booking, facilitating the hospitality business. AI helps to understand occupancy patterns as per holiday season or local events and dynamically adjusts pricing, so marketing efforts result in maximum occupancy and profits [38].

11.4.6 HOTEL OPERATIONS

The front desk is one of the busiest places in a hospitality company; the task can be attending to guests, answering calls, document verification during check-in, query solving, etc. All of which cannot be effectively and efficiently managed by human resources, especially when footfall increases, resulting in poor guest management time and increased effort required for attending all this. AI technology-powered chatbots can handle multiple guests at the front desk and online queries. At the same time, self-check-in kiosks can quicken the check-in process by removing hassles, improving the customer service experience and resulting in guest satisfaction. Having such quick check-in service and consistent guest experience, along with streamlining maintenance workflows and recognizing potential issues, helps Hospitality Company improve service delivery and gain a competitive edge [39].

11.4.7 PERSONALIZATION

AI technology helps to build personal communications and relationships with guests, which are critical for their business to retain customers. AI facilitates personalization, thereby improving service efficiency and effectiveness as personalized communication builds trust and strengthens relationships which helps in improving a loyal customer base. Personalized messages to ask about stay experience, specialized discounts and offers, coupons, etc., help stay connected with loyal customers, facilitating business scalability. By predicting guests' preferences, AI can help to improve productivity and give quick, personalized service. Personalized guest experiences increase loyalty and make them return, which boosts revenue. AI can identify guest preferences and tailor services accordingly, including loyalty rewards programs for customers, enhancing customer satisfaction [40, 41].

11.5 ARTIFICIAL INTELLIGENCE (AI) APPLICATIONS IN HEALTHCARE

Healthcare management had already implemented a lot of AI applications, which are supporting healthcare management in authorization, billing, claims, insurance management, medical research and precision treatment. This data helps to offer more prompted and personalized services for the better experience. The few examples of AI supported application utilized by healthcare industry are discussed in subsections.

11.5.1 PRIOR AUTHORIZATION

AI technology can help in the automation of prior authorizations to reduce the administrative burden of physicians. Hospitals and healthcare providers deal with multiple prior authorizations every day, which involves tasks like identifying patients' health plans, medicines, and drug benefits, identifying prior authorization services as per an individual patient's health plan benefit, and collecting and providing documentation regarding the patient history and medical necessity justification along with referral information for approval, etc., many of these procedure steps can be automated with the use of AI [42].

11.5.2 BILLING AND CLAIMS PROCESSING

Most patients nowadays have medical insurance, and handling claims becomes a tedious, time-consuming, costly task for hospitals and physicians. Billing and insurance costs make up a significant portion of physician and hospital care costs, while claim denial is expensive. AI technology to automate the process and enter and adjust claims helps healthcare organizations streamline the same and make it more reliable through automated verification of claims, saving time, money, and resources for insurance providers and healthcare organizations. AI technology can help to reduce the turnaround time required for regular claims while eliminating repetitive tasks like converting physician notes into standardized codes can be done automatically to make claim procedures faster and more accurate [42].

11.5.3 HEALTHCARE WORKERS PRODUCTIVITY

Healthcare providers have routine, time-consuming tasks like documentation of patient summaries, chart notes, writing prescriptions for medicines and tests orders, etc., which takes time away from their primary focus of providing quality direct patient care while also having the responsibility for maintaining data integrity and privacy of complex electronic health records. AI technology can help healthcare providers eliminate such manual tasks by assisting in administrative workflows, which can save significant time for nurses or physicians and make them more productive. So not only does productivity improve through the use of AI, but it also helps healthcare providers cut administrative costs while making them more efficient in their primary task of providing quality direct patient care.

11.5.4 CLINICAL JUDGMENT AND DIAGNOSIS

AI technology helps in the early detection of diseases, including cancer, as using AI in radiology image analysis and review can speed up the process while providing accuracy. Healthcare organizations can unlock health data for robust diagnosis using cognitive technology. AI can analyze all medical information like case studies of treatment for various diseases, response recovery time, symptoms, medical journal information, etc. It can even use learning algorithms in neural networks to mimic human brains. Such diagnosis helps improve patient care as predictive analytics can support clinical decision-making and take timely preventive actions. AI technology can also be used for pattern recognition, like identifying patients at risk of catching a disease or seeing their health deteriorate due to lifestyle or environmental factors. Moreover, it can help them take early preventive measures to ensure health safety. So be it detecting skin cancer, eye diseases, analyzing 3D retinal scans, etc., AI technology can interpret and infer from data with more reliability and trust. Combining the power of such AI algorithms and diagnosis analytical reports with human pathologists,' physicians' expertise and opinion can help to make successful judgments to detect and diagnose diseases early to provide fast and accurate treatments to patients on a case prioritization basis. Thereby improving care quality and customer experience and satisfaction [43].

11.5.5 PRECISION MEDICINE AND TREATMENT

Perhaps AI's most important application in healthcare is precision medicine. Precision medicine means customizing and providing medical treatment according to each patient's characteristics rather than generalized treatment according to the category of illness or disease. Precision medicine foundation depends on the amounts of medical data collected and stored from multiple digital devices, including data collected from portable health sensors, which patients use everywhere, affordable genome sequencing and medical history found in electronic health records and advanced biotechnology. Using AI technology to analyze such a variety of data helps healthcare providers and medical practices shift from generalized decisions and treatment to more specific personalized treatment decisions. Using cognitive capabilities of physicians and doctors and machine learning algorithms can help provide precision medicine as AI technology can analyze genomic data and detect genetic variations that can increase disease risk early and use precision medicine to treat such illnesses early and accurately, thereby delivering quality patient care. AI can also help medical professionals like clinicians to develop comprehensive disease management plan, coordinate patient care approach and help them manage and make them follow long-term treatment programs for better and complete recovery [44].

11.5.6 MEDICINE RESEARCH AND DRUG DISCOVERY

Multiple drugs begin pre-clinical testing before human testing, and finally, few get human usage approval, which takes years of research and is a significantly costly process. AI technology usage can help in medical research, which can help pharmaceutical companies streamline the drug discovery process and drug repurposing, as AI can help drive oncology drug discovery programs. AI technology can identify various disease causes that were previously unknown to humans and thus can facilitate the testing of medicinal compounds with more accuracy and increase reproducibility. Usage of AI technology in medicinal research and drug discovery can help pharmaceutical companies to reduce the time from clinical testing to human usage for new drugs as well as help them reduce costs and disapproval losses which in turn help patients get new, better, more effective medicines at a lower rate. AI usage in medicinal research and drug discovery can help pharmaceutical companies to shift from the traditional trial and error approach to more patient-driven biology due to the usage of predictive hypotheses, which are derived from medical research data and thus improve the productivity of the pharmaceutical industry's research and development standards [23].

11.6 ARTIFICIAL INTELLIGENCE (AI) IMPACTING MEDICAL TOURISM

AI has redefined medical tourism, as the integration of AI and AI-assisted technologies can offer more accurate patient care, better surgery, timely follow-up, and looking after the safety of hospitals and healthcare workers. Some of the areas where AI and robotics technology are making an impact in the medical tourism industry are discussed in subsections.

11.6.1 BUILDS MEDICAL TOURISM DESTINATION

Handling medical tourists from different countries is difficult for healthcare organizations because of the language barrier. Medical tourists prefer cost savings and value-for-money healthcare services without compromising their primary need for quality treatment and patient care service delivery. Integrating AI technology helps in translation to overcome the communication barrier. Clinical assessment time is reduced as robotics with AI technology can assist doctors or surgeons make faster decisions and, as such, can provide quality treatment to multiple patients. At the same time, robotic-assisted surgery can help deliver accurate treatment faster. Such great diagnosis and treatment help improve the satisfaction levels of medical tourists. Thus, AI and robotics can help healthcare organizations build the capacity to attract more medical tourists and become a medical tourism destination [23, 45].

11.6.2 ROBOTIC AND AI ASSISTED SURGERY

AI technology and robotics-assisted surgery help in performing complex critical operations with precision, thereby ensuring consistency. AI and Robotics assisted critical surgeries like removing tumors can be performed with high accuracy while leaving the organ intact. Medical tourists look for cost-effective treatment for complex and critical medical cases, and healthcare organizations which offer advanced robotics-assisted surgeries and healthcare organizations well recognized for having such infrastructure and facilities become the preferred destination for medical tourism as everyone is aware of medical breakthroughs. Such AI -assisted surgeries with minimum incisions are less painful, and recovery time is faster, which helps the patients get discharged from the hospital faster. Such medical tourists now prefer Robotic and AI-assisted surgery so they can fly back home faster with the satisfaction of receiving proper treatment [23, 46].

11.6.3 PATIENT CARE

Another critical factor for becoming a preferred medical tourism destination for healthcare organizations is improving and adding value to patient care. Caregivers, medical attendants, and nurses must perform multiple repetitive tasks daily like checking vital parameters, which can become a burden when the number of patients increases and can result in human errors fetching false results leading to undesirable consequences. Getting mentally fatigued from such repetitive tasks can become a hindrance and affect patient-healthcare worker interactions, resulting in poor patient care experience for medical tourists. AI technology-powered robots can perform such tasks easily without errors, freeing healthcare workers to attend to more patients with better empathy and improved decision making for understanding patient requirements and providing better patient care resulting in medical tourist service satisfaction [23, 47].

11.6.4 HOSPITAL SAFETY

One of the critical factors for medical tourists in deciding medical tourism destination to avail of healthcare services is the safety standards of healthcare organizations they want to visit. Safety standards are taken into consideration during treatment or surgery, and safety standards throughout hospital facilities. Hospital-acquired infections are a major concern and fear for abroad patients as medical tourism increases the risk of carrying infections back home. Medical tourism destinations, through artificial technology-induced robotic skills, focus on minimal contact to prevent the chances of catching and spreading infections. Whether delivering food, medicines or other supplies, robotic applications focus on preventing frontline health workers from frequent exposure to infections. Since COVID-19, hospital safety standards have increased, and foreign patients have safety issues concerns; thus, AI-induced robotics are the best solutions. Using robotics not only to assist in complex surgeries but also to destroy infection-causing microorganisms better can help disinfect hospitals and clinics. This will help healthcare organizations improve their safety standards and become attractive destinations for medical tourists [48].

11.6.5 ASSISTING HEALTHCARE WORKERS

The Healthcare industry always faces a shortage of medical professionals. With increased patients considering medical tourism growth, there will be a shortage of healthcare workers to assist them. As such,

AI technology-induced robots can fill the gap left by the availability of healthcare workers and help prevent burnout of existing healthcare workers due to work overload. AI technology-induced robots can also help fast recovery for critical care patients who need 24/7 monitoring and assistance, which robots can do efficiently. Tasks like shifting beds and lifting patients, helping in patients' movement on the bed, regularly taking vital parameters and monitoring patient's condition, filling up forms on their behalf, etc., can be done by robotic nurses freeing up medical assistants for providing direct patient care and thus help healthcare organizations handle a large volume of medical tourists [49].

11.6.6 FOLLOW-UPS

One of the main concerns for medical tourists is the lack of follow-up from healthcare organizations after a major treatment or surgery as they fly back home and can't physically visit for routine follow-up check-ups. Having robotics with AI technology-induced can help healthcare organizations monitor and keep track of medical tourists and look after their queries by finding quick solutions to help the patients recover fully at home without needing to visit again physically. AI technology can help doctors or surgeons stay connected with foreign patients through remote patient monitoring, telemedicine, and teleconferencing with their local family doctor. It can assist patients through virtual rehabilitation and thus help improve follow-up services of healthcare organizations, which in turn helps attract and retain medical tourists.

11.7 CONCLUSION

With AI in hospitality companies, the data of guests' stay experience, data related to travel lifecycle and customer satisfaction levels can help the hospitality companies retain and attract customers. AI is revolutionizing the hospitality industry with operations optimization and guest experience streamlining as AI can carry out human functions more effectively and efficiently, thereby driving sales and profits for hospitality companies. Embracing AI technology can help hospitality companies gain a competitive edge by delivering personalized, quick, efficient services to improve guests' experience, leading to customer satisfaction, which is the key in the hospitality industry. AI cannot replace human staff completely. However, can-do repetitive tasks to support them, like by easing the burden of late-night shifts, help reduce employee turnover. Hospitality companies need to adopt and integrate AI into their daily business operations like concierge service, room cleaning, check-in process, etc., to gain a competitive advantage by providing a personalized enhanced guest stay experience.

AI technology usage in the healthcare industry has multiple applications and great benefits for back-end administrative operations and direct patient care. Healthcare workers like nurses and doctors can save significant time using AI for repetitive manual tasks and become more productive and efficient. Using AI technology for analyzing medical data can help improve the quality of patient care and smoothen patient experience as AI assists in administrative workflows. Automating repetitive administration tasks of data collection and storage with the help of AI technology can support doctors and nurses to improve quality efficiency and effectiveness in their primary job of direct patient care. Integrating AI technology in clinical practice can help doctors, physicians, and clinicians focus on tasks that require uniquely human skills and a high cognitive function to deliver better quality patient care.

Data analysis related to pain management, faster accurate diagnosis, cost-efficient treatment and surgery-related discomforts of patients help improve the patient care experience. It is an important integral part of the medical tourism industry, and AI-induced robotics technology can help deliver all this efficiently for healthcare organizations, thereby helping and enabling them to become the preferred destination for medical tourists. Robotics with artificial technology is disrupting and transforming the healthcare sector and is helping medical professionals like doctors and surgeons and helping even associated professionals like nurses deliver better direct patient care. Such improved medical facilities and service quality due to AI and robotics technology revolution in healthcare can boost medical tourism industry growth as global patients can get access to the best-specialized healthcare service from medical professionals anywhere in the world and get the desired treatment and surgery outcome. Medical tourists can avail themselves of such cost-effective services, delivering high value for them. With travel restrictions lifting post COVID pandemic, the medical tourism sector will continue to grow as patients try to get and use the best and most modern healthcare facilities and services available at affordable prices.

KEYWORDS

- **artificial intelligence**
- **drug discovery**
- **healthcare**
- **hospitality industry**
- **hotel industry**
- **hyperdynamic pricing**
- **medical tourism**

REFERENCES

1. Petiwala, F. F., Shukla, V. K., & Vyas, S., (2021). IBM Watson: Redefining artificial intelligence through cognitive computing. In: Prateek, M., Singh, T. P., Choudhury, T., Pandey, H. M., & Gia, N. N., (eds.), *Proceedings of International Conference on Machine Intelligence and Data Science Applications. Algorithms for Intelligent Systems*. Springer, Singapore. https://doi.org/10.1007/978-981-33-4087-9_15.
2. Suhel, S. F., Shukla, V. K., Vyas, S., & Mishra, V. P., (2020). Conversation to automation in banking through chatbot using artificial machine intelligence language. In: *2020 8th International Conference on Reliability, Infocom Technologies and Optimization (Trends and Future Directions) (ICRITO)* (pp. 611–618). doi: 10.1109/ICRITO48877.2020.9197825.
3. Bala, S., Khalid, M. N., Kumar, H., & Shukla, V. K., (2022). The practical enactment of robotics and artificial intelligence technologies in e-commerce. In: *Cyber Intelligence and Information Retrieval* (pp. 455–467). Springer, Singapore.
4. Mathew, D., Shukla, V. K., Chaubey, A., & Dutta, S., (2021). Artificial intelligence: Hope for future or hype by intellectuals. In: *2021 9th International Conference on Reliability, Infocom Technologies and Optimization (Trends and Future Directions) (ICRITO)* (pp. 1–6). doi: 10.1109/ICRITO51393.2021.9596410.
5. Bhattacharya, A., Ghosh, G., Mandal, R., Ghatak, S., Samanta, D., Shukla, V. K., & Mandal, A., (2022). Predictive analysis of the recovery rate from coronavirus (COVID-19). In: *Cyber Intelligence and Information Retrieval* (pp. 309–320). Springer, Singapore.
6. Thange, U., Shukla, V. K., Punhani, R., & Grobbelaar, W., (2021). Analyzing COVID-19 dataset through data mining tool orange. In: *2021 2nd International Conference on Computation, Automation and Knowledge Management (ICCAKM)* (pp. 198–203). doi: 10.1109/ICCAKM50778.2021.9357754.
7. Kappukalar, N. T. F., Shukla, V. K., & Gupta, S., (2021). Automation of disaster recovery and security in cloud computing. In: *2021 International Conference on*

Communication Information and Computing Technology (ICCICT) (pp. 1–6). doi: 10.1109/ICCICT50803.2021.9510110.
8. Thekkethil, M. S., Shukla, V. K., Beena, F., & Chopra, A., (2021). Robotic process automation in banking and finance sector for loan processing and fraud detection. In: *2021 9th International Conference on Reliability, Infocom Technologies and Optimization (Trends and Future Directions) (ICRITO)* (pp. 1–6). doi: 10.1109/ICRITO51393.2021.9596076.
9. Petiwala, F. F., Shukla, V. K., Mishra, V. P., & Saini, S., (2021). Smart parking system through automation in license plate recognition. In: *2021 9th International Conference on Reliability, Infocom Technologies and Optimization (Trends and Future Directions) (ICRITO)* (pp. 1–7). doi: 10.1109/ICRITO51393.2021.9596554.
10. Zaidi, S. F. N., Shukla, V. K., Mishra, V. P., & Singh, B., (2021). Redefining home automation through voice recognition system. In: Hassanien, A. E., Bhattacharyya, S., Chakrabati, S., Bhattacharya, A., & Dutta, S., (eds.), *Emerging Technologies in Data Mining and Information Security: Advances in Intelligent Systems and Computing* (Vol. 1300). Springer, Singapore. https://doi.org/10.1007/978-981-33-4367-2_16.
11. Gupta, N., Sharma, P., Deep, V., & Shukla, V. K., (2020). Automated attendance system using openCV. In: *2020 8th International Conference on Reliability, Infocom Technologies and Optimization (Trends and Future Directions) (ICRITO)* (pp. 1226–1230). doi: 10.1109/ICRITO48877.2020.9197936.
12. Nam, K., Dutt, C. S., Chathoth, P., Daghfous, A., & Khan, M. S., (2021). The adoption of artificial intelligence and robotics in the hotel industry: Prospects and challenges. *Electronic Markets, 31*(3), 553–574.
13. Mohammed, U. S., Shukla, V. K., Sharma, R., & Verma, A., (2022). Next step to the future of restaurants through artificial intelligence and facial recognition. In: *Cyber Intelligence and Information Retrieval* (pp. 89–101). Springer, Singapore.
14. Albuquerque, D. D., Shukla, V. K., Verma, A., Tyagi, S. K., & Sharma, P., (2020). Enhancing sustainable customer dining experience through QR code and geo-fencing. In: *2020 International Conference on Computation, Automation and Knowledge Management (ICCAKM)* (pp. 190–196). doi: 10.1109/ICCAKM46823.2020.9051470.
15. Bhardwaj, M., Pandey, N., Shukla, V. K., Singh, A. V., & Gupta, N., (2021). Review and analysis of security model in healthcare system. In: *2021 9th International Conference on Reliability, Infocom Technologies and Optimization (Trends and Future Directions) (ICRITO)* (pp. 1–7). doi: 10.1109/ICRITO51393.2021.9596090.
16. Trayush, T., Bathla, R., Saini, S., & Shukla, V. K., (2021). IoT in healthcare: Challenges, benefits, applications, and opportunities. In: *2021 International Conference on Advance Computing and Innovative Technologies in Engineering (ICACITE)* (pp. 107–111). doi: 10.1109/ICACITE51222.2021.9404583.
17. AltexSoft., (2022). *AI in Healthcare*. [online] Available at: https://www.altexsoft.com/ai-solutions-for-healthcare/ (accessed on 07 September 2023).
18. Murlidharan, S., Shukla, V. K., & Chaubey, A., (2021). Application of machine learning in precision agriculture using IoT. In: *2021 2nd International Conference on Intelligent Engineering and Management (ICIEM)* (pp. 34–39). doi: 10.1109/ICIEM51511.2021.9445312.
19. Tamang, M. D., Kumar, S. V., Anwar, S., & Punhani, R., (2021). Improving business intelligence through machine learning algorithms. In: *2021 2nd International*

Conference on Intelligent Engineering and Management (ICIEM) (pp. 63–68). doi: 10.1109/ICIEM51511.2021.9445344.

20. Dasgupta, A., Mishra, V. P., Jha, S., Singh, B., & Shukla, V. K., (2021). Predicting the likelihood of survival of Titanic's passengers by machine learning. In: *2021 International Conference on Computational Intelligence and Knowledge Economy (ICCIKE)* (pp. 52–57). doi: 10.1109/ICCIKE51210.2021.9410757.
21. Wang, W., & Siau, K., (2018). Ethical and moral issues with AI: A case study on healthcare robots. In: *Twenty-fourth Americas Conference on Information Systems, New Orleans, LA* (Vol. 16, p. 2019).
22. ForeSee Medical., (2022). *Artificial Intelligence (AI) In Healthcare & Hospitals.* [online] Available at: https://www.foreseemed.com/artificial-intelligence-in-healthcare (accessed on 07 September 2023).
23. 2022. [online] Available at: https://drprem.com/medical-tourism/robotics-with-intelligence-would-be-a-game-changer-in-medical-tourism-sector/ (accessed on 07 September 2023).
24. AltexSoft., (2022). *AI in Healthcare.* [online] Available at: https://www.altexsoft.com/ai-solutions-for-healthcare/ (accessed on 07 September 2023).
25. Hassan, V., & Bellos, G. (2022). COVID-19: Reshaping Medical Tourism through Artificial Intelligence (AI) and Robotics. *Athens Journal of Tourism*, *9*(2), 77–98.
26. Al-Talabani, H., Kilic, H., Ozturen, A., & Qasim, S. O., (2019). Advancing medical tourism in the United Arab Emirates: Toward a sustainable health care system. *Sustainability*, *11*(1), 230.
27. Ile, F. L., & Ţigu, G., (2017). Medical tourism market trends-an exploratory research. In: *Proceedings of the International Conference on Business Excellence* (Vol. 11, No. 1, pp. 1111–1121).
28. Shukla, V. K., & Verma, A., (2019). Enhancing LMS experience through AIML base and retrieval base chatbot using R language. In: *2019 International Conference on Automation, Computational and Technology Management (ICACTM)* (pp. 561–567). doi: 10.1109/ICACTM.2019.8776684.
29. Athota, L., Shukla, V. K., Pandey, N., & Rana, A., (2020). Chatbot for healthcare system using artificial intelligence. In: *2020 8th International Conference on Reliability, Infocom Technologies and Optimization (Trends and Future Directions) (ICRITO)* (pp. 619–622). doi: 10.1109/ICRITO48877.2020.9197833.
30. Drexler, N., & Lapré, V. B., (2019). For better or for worse: Shaping the hospitality industry through robotics and artificial intelligence. *Research in Hospitality Management*, *9*(2), 117–120.
31. Punhani, R., Arora, V. P. S., Sabitha, A. S., & Shukla, V. K., (2021). Segmenting e-commerce customer through data mining techniques. In: *Journal of Physics: Conference Series* (Vol. 1714, No. 1, p. 012026). IOP Publishing.
32. Dubey, S. K., Mittal, S., Chattani, S., & Shukla, V. K., (2021). Comparative analysis of market basket analysis through data mining techniques. In: *2021 International Conference on Computational Intelligence and Knowledge Economy (ICCIKE)* (pp. 239–243). doi: 10.1109/ICCIKE51210.2021.9410737.
33. Beena, F., Mearaj, I., Shukla, V. K., & Anwar, S., (2021). Mitigating financial fraud using data science: A case study on credit card frauds. In: *2021 International*

Conference on Innovative Practices in Technology and Management (ICIPTM) (pp. 38–43). doi: 10.1109/ICIPTM52218.2021.9388345.
34. Mehrotra, A., & Kumar, M. P., (2020). Artificial intelligence in hospital administration. *Journal of Dental and Orofacial Research, 16*(1), 47–50.
35. Jhamb, S., & Shukla, V. K., (2020). The role of teacher and technology in foreign language acquisition. *IUP Journal of English Studies, 15*(3), 7–14.
36. Jhamb, S., Gupta, R., Shukla, V. K., Mearaj, I., & Agarwal, P., (2020). Understanding complexity in language learning through data visualization using Python. In: *2020 International Conference on Computation, Automation and Knowledge Management (ICCAKM)* (pp. 268–274). doi: 10.1109/ICCAKM46823.2020.9051512.
37. Zaidi, S. F. N., Shukla, V. K., Mishra, V. P., & Singh, B., (2021). Redefining home automation through voice recognition system. In: Hassanien, A. E., Bhattacharyya, S., Chakrabati, S., Bhattacharya, A., & Dutta, S., (eds.), *Emerging Technologies in Data Mining and Information Security: Advances in Intelligent Systems and Computing* (Vol 1300). Springer, Singapore. https://doi.org/10.1007/978-981-33-4367-2_16.
38. Ruel, H., & Njoku, E., (2020). AI redefining the hospitality industry. *Journal of Tourism Futures.*
39. Liu, Y., & Zhang, N., (2021). Study of hotel AI application and development trends in 5G Era. In: *2021 4th International Conference on Advanced Electronic Materials, Computers and Software Engineering (AEMCSE)* (pp. 1051–1054). IEEE.
40. Nawazish, H. N., & Shukla, V. K., (2020). Integrating "random forest" with indexing and query processing for personalized search. In: *2020 8th International Conference on Reliability, Infocom Technologies and Optimization (Trends and Future Directions) (ICRITO)* (pp. 280–283). doi: 10.1109/ICRITO48877.2020.9197888.
41. Siraj, A., & Shukla, V. K., (2020). Framework for personalized car parking system using proximity sensor. In: *2020 8th International Conference on Reliability, Infocom Technologies and Optimization (Trends and Future Directions) (ICRITO)* (pp. 198–202). doi: 10.1109/ICRITO48877.2020.9197853.
42. Evicore.com. (2022). *How Artificial Intelligence Can Make Hospital Administration More Efficient.* [online] Available at: https://www.evicore.com/insights/how-artificial-intelligence-can-make-hospital-administration-more-efficient (accessed on 07 September 2023).
43. ApexTech., (2022). *6 Applications of AI Technology in the Hospitality Industry* | ApexTech. [online] Available at: https://apextechinc.com/6-applications-of-ai-technology-in-the-hospitality-industry/ (accessed on 07 September 2023).
44. HIMSS, (2022). *AI in Healthcare: How It's Changing the Industry.* [online] Available at: https://www.himss.org/resources/ai-healthcare-how-its-changing-industry (accessed on 07 September 2023).
45. Zolfagharian, M., Rajamma, R. K., Naderi, I., & Torkzadeh, S., (2018). Determinants of medical tourism destination selection process. *Journal of Hospitality Marketing & Management, 27*(7), 775–794.
46. Bodenstedt, S., Wagner, M., Müller-Stich, B. P., Weitz, J., & Speidel, S., (2020). Artificial intelligence-assisted surgery: Potential and challenges. *Visceral Medicine, 36*(6), 450–455.
47. Lunt, N., & Carrera, P., (2010). Medical tourism: Assessing the evidence on treatment abroad. *Maturitas, 66*(1), 27–32.

48. Furrow, B. R., (2020). The limits of current AI in health care: Patient safety policing in hospitals. *NEULR*, *12*, 1.
49. Lai, Y., Kankanhalli, A., & Ong, D. (2021). Human-AI collaboration in healthcare: A review and research agenda, Proceedings of the 54th Hawaii International Conference on System Sciences, Page 390–399, Kauai, Hawaii, USA, January 5, 2021, https://hdl.handle.net/10125/70657 (accessed on 07 September 2023).

CHAPTER 12

Leveraging Potentials of Artificial Intelligence and Robotics for the Tourism Industry: A Study

SONALI VYAS,[1] SHAURYA GUPTA,[1] and VINOD KUMAR SHUKLA[2]

[1]*School of Computer Science, University of Petroleum and Energy Studies, Dehradun, Uttarakhand, India*

[2]*Department of Engineering and Architecture, Amity University, Dubai, UAE*

ABSTRACT

Travel, tourism, and hospitality sectors also benefit from artificial intelligence (AI) and IoT technologies. Its applications can be seen in varied tour and travel websites, chat robots, smart tour and travel agents, and predicting systems. Recent advancements in Big Data and computational intelligence have led to significant enhancements in the field of AI. With time, the whole scenario, including all processes and procedures of the travel and tourism industry, has changed. Information and Communication technologies in tourism have coined the concept of e-tourism, which has opened a new plethora of opportunities in contemporary tourism and hospitality business. It has enabled researchers in developing models to read the behavior of travelers besides suggesting them varied holiday packages, apart from that handling and managing data generated from travelers and the hospitality sector end. According to information and communication technology, AI is considered as next level in the travel and tourism business. An AI-based system is responsible for sensing exterior info, comprehending it, and then acting in order to accomplish said objectives. The manuscript discusses the foundations of IT in AI, which is quite pertinent for travel besides vacation business apart from addressing AI tools and systems applicable in this particular segment. Secondly, it discourses the impact of the pandemic on the travel and tourism business and its proposed solutions apart from the role of AI, its influence on the existing hospitality business, and the role of chatbots and robots with some discussion on the existing AI framework implementation for tourism business. Lastly, it is concluded with the introduction of Metaverse methods for uplifting tourism with future advancements.

12.1 INTRODUCTION

Artificial intelligence (AI) is significant in travel besides tourism for numerous motives [1]. Travelers must make various conclusions regarding upcoming journeys, including selecting a terminus, conveyance, and lodging. Choices have a noteworthy inspiration on vacationers' contentment with their trip. These agents of these service sector face a comparable encounter while finding the finest match amongst clients besides tourism packages which is quite customized according to their needs [2]. As sightseers reach a certain destination, they should be guided by varied travel bots regarding the nearest sightseeing places apart from food joints or any places of cultural heritage ages [3]. The usage of AI can presently be entrenched in the case of information dispensation arrangements in real-time surroundings. Figure 12.1 shows various AI components for the tourism business.

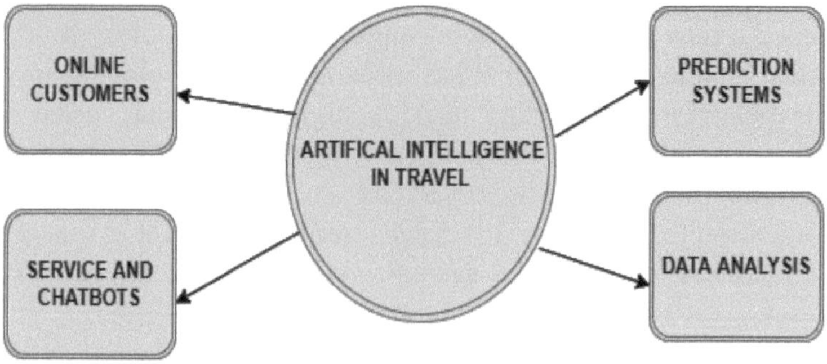

FIGURE 12.1 AI components in travel and tourism.

12.1.1 PANDEMIC IMPACT ON TRAVEL AND TOURISM

Travel and tourism stand as one of the ecosphere's chief monetary segments. In the year 2019, it accounted for 7% of worldwide trade. For many countries, tourism represents about 20% of their GDP. Due to the COVID-19 endemic, the tourism sector was the most affected segment, impacting economies, living, and communal amenities besides development prospects across the globe [4]. Travel and tourism provide about one out of 10 jobs besides living for numerous livelihoods in emerging and urbanized economies. The travel and tourism sectors for small islands account for almost 80% of exports.

12.1.2 CONTEMPORARY ROLE OF ARTIFICIAL INTELLIGENCE (AI)

AI structures in this service sector may be a self-sustaining system or entrenched in prevailing applications, and it comprises recommender structures, personalization systems, chatty or conversational systems like voice-based assistants or travel bots, foretelling gears, independent representatives, linguistic support applications besides, smart leisure industry terminuses [5]. Generally, travelers will be interacting with technologies that integrate many systems. For instance, a lodger may communicate with a robot, which is already an integrated entity of conversational and recommender systems [6].

12.1.3 ROLE OF TRAVEL BOT IN TRAVEL AND TOURISM INDUSTRY

Whereas communication with any end-user is considered, it is always facilitated with the help of a voice assistant or chatbot [7, 8]. For the past few years, travelers, and tourists used to elect any terminuses to be visited, with the help of catalog pictures. With the advent of the Internet, information is available in a real-time and dynamic mode, which has helped sightseers in making improved choices. The transformations in the field of AI have allowed sightseers to get the best choices which outfit them in all ways besides permitting businesses to experience end users' precise necessities [9, 10].

12.2 LITERATURE REVIEW

Many tourism organizations rely mostly on technological advances in the late 20th and early 20th centuries. They offer their clients facilities and products with these applications like cyberspaces, social media, mobile apps, augmented reality (AR), VR, mixed reality (MR), robots/chatbots, and self-help kiosks [11]. For numerous years, visitors had to choose destinations, tourist places, and details of different events by using catalog pictures. AI aids users in finding improved and relevant information by giving them better flexibility, advancing decision-making and delivering a healthier tourism experience. Nevertheless, even with this extra information, the price has become a crucial factor in decision-making. AI transforms this behavior by permitting tourists to discover substitutes that suit them best and businesses to adapt this to their customers' precise requirements. It does this via personalization methods and reference systems [12]. AI is used in tourism for various purposes, such as refining the personalization levels, modifying customers' recommendations, and ensuring fast response time even in the absence of staff. AI has become so significant in the industrial environment that it is utilized to support and connect with customers, thereby strengthening the quality of interaction [13]. Recent visitors are more challenging than always, demanding immediate replies on online stages, and AI enables industries to meet the demand for fast response time. Additionally, chatbots and AI-powered traveling bots permit for constant quality of the product. AI is not that common in the tourism industry yet, but the expectation is that it will be used more frequently to improve the quality of products and services [2]. Along with AI, IoT is also playing a very vital role in embracing the experience of tourist [15, 16].

12.3 APPLICATION OF ARTIFICIAL INTELLIGENCE (AI) FOR TOURISM

AI is one of the most advanced developments that has transformed numerous businesses worldwide. It can be expressed as the advancement of computer systems that can accomplish activities requiring the intelligence of the human mind. The tourism industry's AI systems can be considered impartial embedded systems in prevailing applications. Such systems comprise personalization and recommendation systems, prediction tools, autonomous agents, conversational systems (like voice assistants and chatbots), and smart tourism destinations. Even though we assess every system distinctly, it should be noted that tourists will often communicate with technologies that combine such systems. Smart destinations are systems shaped in destinations where these advanced technologies are organized, along with additional social and organizational structures. Figure 12.2 shows the structure of AI-enabled devices and integrated systems.

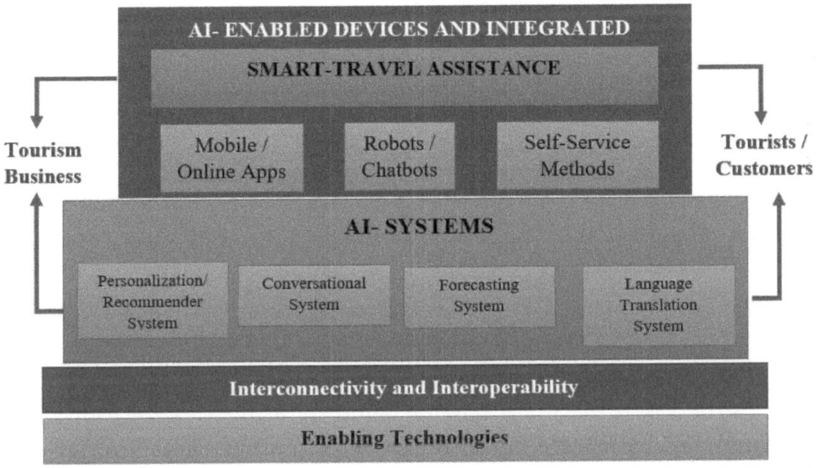

FIGURE 12.2 AI-systems and technologies for tourism business.

12.4 AI-ENABLED DEVICES AND INTEGRATED SYSTEMS

12.4.1 PERSONALIZATION AND RECOMMENDER SYSTEMS

As the number of options available to users increases exponentially with online environments, the use of recommendation systems in tourism has become more critical. Typically, recommendation systems match the properties of available opportunities with user profiles to make recommendations about the most suitable options [17]. Personalization methods offer users customized information based on their requirements and constraints. Therefore, personalization procedures mean corporations move from too much marketing to single marketing. Personalization practices entail a huge quantity of information about the behavior of the user in order to define an accurate profile.

12.4.2 CHATBOTS AND VOICE ASSISTANTS

Speech systems are occasionally called chatbots or virtual agents. They include technologies like NLP and speech acknowledgment and are presently pervasive. For instance, they are present as personal assistants on in-home speakers and smartphones (with saleable structures like Google Assistant, Amazon Alexa, Microsoft Cortana and Apple Siri) and text-based chatbots incorporated in websites and kiosks [18].

12.4.3 FORECASTING

In the tourism industry, forecasting is utilized for understanding the demand of users, designing marketing plans, allocating financial and human resources, identifying fraud in restaurants, and assisting the management by providing facilities and fulfilling maintenance needs.

12.4.4 LANGUAGE TRANSLATION APPLICATIONS

It is identified that the main barrier faced by almost every tourist is the local language while traveling. It results in anxiety and discomfort for the tourists. Since personalization can assist the tourists in finding new tourist spots, auto-translation can make it easier for tourists to navigate the destination, permitting them to discover and participate in all kinds of activities. AI integrated with NLP and ML helps develop auto-translation applications and concurrent translation systems.

12.4.5 ROBOTS

A robot is an independent machine comprising AI and sensor technology, which lets the robot decide and accomplish activities. Such AI-powered professional service robots are used to rationalize procedures and improve activities which are traditionally accomplished by front desk personnel. The service robots can deal with many human inadequacies in tourism, like language issues and labor scarcities.

12.4.6 SMART TRAVEL ASSISTANTS

As AI, NLP, mobile devices and speech recognition are improving, the idea of smart travel assistants is also gaining interest and viability. Such applications having assistants are accustomed to the user (i.e., their predilections and convenience) and can therefore unconventionally make recommendations by anticipating users' requirements. These systems are smart concierges, autonomous agents, and smart travel agents. A smart travel assistant would be able to integrate numerous services at one destination, handle place and time constraints and discover appropriate ways to get the tourist travel within an anticipated budget.

12.5 SMART TOURISM AND SMART DESTINATIONS

Smart tourism can be well-defined as tourism reinforced by joint efforts in the field of collecting and integrating data from visual infrastructure, social media, government/organization resources and human bodies/minds, as well as the use of advanced technology to convert that data into the real scenario and business value proposals that focus on competence, enrichment, and sustainability. Thus, AI plays an important role in the deployment and development of smart tourism, as the conversion of data into experience and value propositions will be vested by AI. Smart destinations and tours are natural digital systems where AI is key actor. However, to improve these systems, many varied organizational components need to integrate and communicate with one another [19].

12.6 UTILIZATION OF ROBOTS AND CHATBOTS IN TOURISM

Chatbots are computer programs that promote human conversation using natural language. The visitor's need for consistent sources of information is continuously growing, so chatbots are planned to meet the tourist's needs. Travel chatbots improve the travel experience as they direct visitors through all parts of their travels, from ticket booking to common travel assistance. Travel chatbots are categorized as Facebook, customer service, and AI-enabled travel bots.

Robots can be categorized as service robots and industrial robots. Industrial robots are utilized for industrial activities, like soldering, packaging, repairing, and other processes in production and manufacturing. Service robots are programmed to provision for people through social communication. Service robots can be divided into personal and professional service robots. Corporations employ a proficient service, and people use personal provision to perform non-commercial activities. As an integral part of the service sector, the tourism sector has embraced the usage of professional service robots. The segment is fortified with automated service technology and robotic technology that covers a wide range of operations [20]. The acceptance of robots has encouraged advanced production and financial development. Furthermore, robots have arisen as a prevalent technology because the concept of automation and self-help plays an important role in the client experience. The utilization of robots directs to enhanced speediness, lower-cost efficiency and accurateness [21].

12.7 METAVERSE-ENABLED TOURISM

The metaverse perspective and related sensations have profoundly changed people's perceptions of the imminent technology and its potential impression on the tourism industry. Industries are keen to discover ways to monetize metaverse and generate virtual hospitality and information on tourism and services for their users. Powered by Web 3.0, metaverse denotes a 3-D virtual environment focused on social networking. Aside from its ability to endorse social media, metaverse generates a sense of belonging, which is why it considers the tourism industry. It is important for tourism business people to not only identify such a new situation but also make plans to enthusiastically contribute to this innovative world and utilize this prospect to create a sustainable competitive gain. While metaverse will not substitute the experience of real-life tourism, it will surely change the way customers use travel products and services [14].

12.8 CONCLUSION AND FUTURE SCOPE

This chapter discusses the travel and tourism sector and the efficient role of AI in it apart from the scenario of the hospitality industry after a pandemic, besides the role of travel bots in assisting end-users or customers in planning their holiday destinations. Still, there's a lot of scope wherein AI can be deployed for enhanced and amended services in the tour and travel industry. With the help of AI applications, any hotel suite or room can be transformed into a desirable sightseer spot. Travelers will have the choice of transforming their complete chamber into their preferred terminus spot, i.e., the area can be transformed into a simulated 3-D creation resembling their preferred sightseer spot, and the traveler can experience the same view from their suite or room after that they can narrow down to a destination of their choice. This chapter also discusses the upcoming trend of using robots and chatbots to guide hotel visitors to rooms or buffet areas, hold their baggage, and provide housekeeping services. This also introduces the forthcoming era of a metaverse in tourism, which will provide or deliver a live view of real-world developments in front of explorers like shops, guesthouses, and shopping malls, thereby facilitating location-based proficiencies and a virtual tour concept for providing an enhanced experience to the tourists.

KEYWORDS

- **artificial intelligence**
- **chatbots**
- **integrated systems**
- **language translation**
- **metaverse**
- **robotics**
- **smart travel assistants**
- **tourism**

REFERENCES

1. Zsarnoczky, M., (2017). How does artificial intelligence affect the tourism industry? *VADYBA, 31*(2), 85–90.
2. Popesku, J., (2019). Current applications of artificial intelligence in tourism and hospitality. In: *Sinteza 2019-International Scientific Conference on Information Technology and Data Related Research* (pp. 84–90). Singidunum University.

3. Tussyadiah, I., (2020). A review of research into automation in tourism: Launching the annals of tourism research curated collection on artificial intelligence and robotics in tourism. *Annals of Tourism Research, 81*, 102883.
4. Chang, D. S., & Wu, W. D., (2021). Impact of the COVID-19 pandemic on the tourism industry: Applying TRIZ and DEMATEL to construct a decision-making model. *Sustainability, 13*(14), 7610.
5. Kirtil, I. G., & Aşkun, V., (2021). Artificial intelligence in tourism: A review and bibliometrics research. *Advances in Hospitality and Tourism Research (AHTR).*
6. Solakis, K., Katsoni, V., Mahmoud, A. B., & Grigoriou, N., (2022). Factors affecting value co-creation through artificial intelligence in tourism: A general literature review. *Journal of Tourism Futures*, (ahead-of-print).
7. Yang, L., Henthorne, T. L., & George, B., (2020). Artificial intelligence and robotics technology in the hospitality industry: Current applications and future trends. *Digital Transformation in Business and Society*, 211–228.
8. Jabeen, F., Al Zaidi, S., & Al Dhaheri, M. H., (2021). Automation and artificial intelligence in hospitality and tourism. *Tourism Review*.
9. Lv, H., Shi, S., & Gursoy, D., (2022). A look back and a leap forward: A review and synthesis of big data and artificial intelligence literature in hospitality and tourism. *Journal of Hospitality Marketing & Management, 31*(2), 145–175.
10. Lee, M., Kwon, W., & Back, K. J., (2021). Artificial intelligence for hospitality big data analytics: Developing a prediction model of restaurant review helpfulness for customer decision-making. *International Journal of Contemporary Hospitality Management*.
11. Touni, R., (2020). The application of robots, artificial intelligence, and service automation in the Egyptian tourism and hospitality sector (possibilities, obstacles, pros, and cons). *Journal of Association of Arab Universities for Tourism and Hospitality, 19*(3), 269–290.
12. Bulchand-Gidumal, J. (2022). Impact of artificial intelligence in travel, tourism, and hospitality. In: *Handbook of e-Tourism* (pp. 1943–1962). Cham: Springer International Publishing.
13. Samala, N., Katkam, B. S., Bellamkonda, R. S., & Rodriguez, R. V., (2020). Impact of AI and robotics in the tourism sector: A critical insight. *Journal of Tourism Futures.*
14. Gursoy, D., Malodia, S., & Dhir, A., (2022). The metaverse in the hospitality and tourism industry: An overview of current trends and future research directions. *Journal of Hospitality Marketing & Management*, 1–8.
15. Verma, A., & Shukla, V., (2019). Analyzing the influence of IoT in tourism industry. In: *Proceedings of International Conference on Sustainable Computing in Science, Technology and Management (SUSCOM).* Amity University Rajasthan, Jaipur-India.
16. Verma, A., Shukla, V. K., & Sharma, R., (2021). Convergence of IoT in tourism industry: A pragmatic analysis. In: *Journal of Physics: Conference Series* (Vol. 1714, No. 1, p. 012037). IOP Publishing.
17. Kbaier, M. E. B. H., Masri, H., & Krichen, S., (2017). A personalized hybrid tourism recommender system. In: *2017 IEEE/ACS 14th International Conference on Computer Systems and Applications (AICCSA)* (pp. 244–250). IEEE.

18. Calvaresi, D., Ibrahim, A., Calbimonte, J. P., Schegg, R., Fragniere, E., & Schumacher, M., (2021). The evolution of chatbots in tourism: A systematic literature review. *Information and Communication Technologies in Tourism, 2021*, 3–16.
19. Kulualp, H. G., & Sarı, Ö., (2020). Smart tourism, smart cities, and smart destinations as knowledge management tools. In: *Handbook of Research on Smart Technology Applications in the Tourism Industry* (pp. 371–390). IGI Global.
20. Cain, L. N., Thomas, J. H., & Alonso, Jr. M., (2019). From sci-fi to sci-fact: The state of robotics and AI in the hospitality industry. *Journal of Hospitality and Tourism Technology*.
21. Croce, N. (2018). Service robotics in hospitality: developing business model guidelines for a service robotics startup by understanding what is driving and influencing adoption of robots in hotels., Document type: Master Thesis, Rights: Nicola Croce.

CHAPTER 13

Mechanical AI and Customer Perception: A Study on Hospitality Aspirants

VARINDER SINGH RANA[1] and ASHISH RAINA[2]

[1]*Department of Hospitality and Tourism Management, City University College of Ajman, UAE*

[2]*Kanya Maha Vidyalaya, Jalandhar, Punjab, India*

ABSTRACT

Automation has taken hold in nearly every industry in the current millennium. The technological advancements in the field of mechanics with advanced artificial intelligence (AI) systems of optimum efficiency have proved to stand side by side with the human workforce. The hospitality and tourism industry has also benefitted the business by adopting the technology of delivery robots, automated concierge, conveyor-based eateries, self-service check-in and check-out kiosks, chatbots, and other mechanical AI systems. The chapter presents a study surveyed the perception of hospitality and tourism aspirants about the use of mechanical AI in the various aspects of the hospitality and tourism industry. The data is collected with purposive sampling by circulating online questionnaires in the closed hospitality and tourism student groups. The study recommends that the aspirants in the tourism and hospitality sector have a favorable perception of the use of mechanical AI in the hospitality and tourism industry. Also, the aspirants believe that mechanical AI can add more value to the service industry, but it cannot replace humans in the hospitality and

Artificial Intelligence for Smart Technology in the Hospitality and Tourism Industry.
Vinod Kumar Shukla, Amit Verma, and Jean Paolo G. Lacap (Eds.)

tourism industry. The chapter recommends that the hospitality and tourism sector balance the activities of the human workforce and AI systems so that these systems can work on monotonous tasks with proficiency whereas the human workforce keeps on delivering warm, personalized services to the guests.

13.1 INTRODUCTION

Customer satisfaction is the objective and rationale of any enterprise. In the context of hospitality, visitor contentment has traditionally been defined as the degree of total pleasure or happiness felt by consumers due to the capacity of experience to meet their demands, expectations, and requirements [1]. Humans need food to survive, and it's widely accepted that the best result of farming is the variety of foodstuffs that farmers create and distribute. Agricultural goods are essential to the economic progress of every country. Economic growth in the United States and worldwide has greatly influenced the agriculture and the food sector as a whole. The food sector relies on high-quality goods and safe delivery to meet this demand [2]. When it comes to greeting and entertaining visitors/guests, the term hospitality is considered [3]. Hospitality services are mostly focused on providing F&Bs and lodging to guests and visitors, which can be offered in commercial and non-commercial places [4]. The hospitality sector's fundamental virtue is the guest's service and the relationship created between the guests and the host. The industry is all about the service experiences created for the guests by the hosts [5]. Service and quality demands have risen significantly in recent decades, along with the overall character of the hospitality industry. There might be a significant disconnect between what customers expect and what they receive [6]. Around 80% of consumers expect a great experience, yet just 8% are happy with the reality, according to their poll. According to various circumstances, such as the extensive availability and dissemination of IT, visitor satisfaction has evolved. Today's tech-savvy consumers expect technology-driven products and customized experiences [7]. A growing number of hospitality businesses are adopting new methods of service experience management in response to customer expectations and progressing in the guest service experience. These new methods include service automation, mobile applications, AI, and even service robots [8].

The term "artificial intelligence" (AI) has typically been used to describe a machine capable of learning, planning, perceiving, and processing natural language [9]. Theory and development of computers capable of doing activities ordinarily requiring human intellect such as visual perception, speech recognition, decision-making, and language translation [10]. In computer science, AI is a subfield that focuses mostly on designing machines designed to behave like humans. Inventor of AI, John McCarthy, described AI as "the science and engineering of constructing intelligent machines, especially clever computer programs" [11]. Two of the most popular AI algorithms are called machine learning and deep learning. The people, businesses, and government organizations rely on these models to help them make predictions based on historical data. In the food business, machine learning models are being created to handle the complexity and variety of data. AI has made significant progress during the past few decades. As a result, research into AI-based smart agriculture and the sophisticated food sector is very necessary. In order to meet the needs of society, these methods ensure that high-quality products are delivered on time. Using this cutting-edge technology, the food sector can produce more food in less time, boosting profits by orders of magnitude [12]. The International Federation of Robotics (2018) reported that a total of 109,543 service robots were sold globally in 2017, an 85% increase over 2016. Sales of professional service robots were expected to reach 736,000 units in 2021 but were affected by the COVID pandemic [13]. Robotics and automation have spread to a wide range of industries, including the hospitality business, where robots are used in a variety of ways in a variety of sectors [3, 8]. The Robotics, Artificial Intelligence, and Service Automation Association (RAISA), published in reputable academic outlets prior to 2019, shows that restaurants are the most interested in the topic than any other part of the hospitality and tourism industry [14].

13.2 LITERATURE REVIEW

Many firms, especially the hotel and tourism industries, can "transform" their operations using machine learning, resulting in lower costs, increased production, and improved efficiency and dependability. It is not new for airports such as passenger services and e-passport gates to employ service automation and, in particular, self-service technology, as well as for hotels such as check-in/out or travel agencies to give information. AI has gained speed in recent years due to the rising importance of facial recognition technology, which allows for various analytical and research methods to be developed. Many firms, especially those in the hotel and tourism industries, can "transform" their operations using AI, resulting in lower costs, increased production, and improved efficiency and dependability. It is not new for airports such as passenger services and e-passport gates to employ service automation, in particular, self-service technology, and for hotels to offer self-check-in/out. Travel agencies also provide information through automated kiosks [15].

Compared to other industries (such as engineering), machine learning-based robotics in the hotel and tourism sectors have only recently been introduced. However, they are already being used more often since they offer an affordable solution to many complicated challenges creating better customer experiences [8].

Burgess, 1985 defined hospitality as a social connection characterized by cordial, inviting, and warm behavior toward consumers [16]. Customer-facing personnel and their capacity to create an enjoyable, relaxing, and "hospitable" atmosphere is viewed as the foundation of service quality in this view. The quality of the service process is therefore defined as the communication amongst clients and the staff responsible for handling the guest queries directly. Sipe & Testa [17] published that the quality of a hotel is no longer determined by the interactions between "hosts and guests" but rather by the architecture, technology, and amenities provided in the rooms [17]. The evolution of hospitality goods and services has been driven by the need to provide guests with both a "service" and a "memorable experience." There are numerous advantages and disadvantages of using mechanical AI over human employees, as follows:

- Ability to work 24 hours a day;
- Implement the work correctly and on time;
- To expand the scope of the tasks through software and hardware upgrades;
- Predictable service quality.

Mechanical AI does not get bored and can perform the same tasks repeatedly without complaint; these also do not go on strikes or get ill. However, mechanical AI can only function in a structured environment for a short time and cannot perform jobs that have not been taught or for which the necessary technology is lacking. Also lacking originality and a human approach, mechanical AI is emotionless when engaging with visitors and staff [18].

There are endless mechanical AI applications in tourism and hospitality. For example, these service devices can be used to move objects (e.g., by hand or by the robot), provide information (e.g., by scanning bar codes), print documents, take a picture of a guest (e.g., by shooting a selfie), clean the floor, disinfect the premises or flip burgers. As a result, research suggests that robots are most suited for monotonous, unclean, boring, or hazardous jobs that people would rather avoid. Mechanical AI duties like supplying information, housekeeping, cleaning, waste collection, baggage carrying, and document processing are most suited to the tourist sector specialists [8]. Machine-delivered massages, childcare, hairdressing, dancing with visitors, and other tasks that subordinate a human to an AI machine, either physically or emotionally, were not deemed suitable.

AI and personnel service quality explain major differences in total service quality, customer satisfaction, and loyalty. Only specific service quality parameters affected desired outcomes. AI becomes unfavorable and negligible when regressed against employee service quality.

According to findings, social impact is stronger in hotel services than in airline services. Airline passengers demand more from AI than hotel guests. Tourism services are less likely to employ AI than airline services [19]. In a study aimed to examine the impact of the perceived utility of hotel technology, such as AI, robots, and service automation, on consumers' behavioral intentions, the author published that the technological advancements offered in the hotels have a stronger behavior intention by the regular hotels guests ad compared to the first-time users [20].

13.3 RESEARCH METHODOLOGY

After reviewing the literature, a structured questionnaire was developed to collect the data. The data was collected between December 2021 and February 2022. The link to an online questionnaire was emailed to the student database and put in restricted social media groups of hospitality students from various institutions. This approach guaranteed that only eligible respondents were contacted and had access to the survey. A total of 306 complete responses were recorded during the survey. The questionnaire has three sections. The first section gathered information on the demographic characteristics of respondents. The second section questioned if respondents are aware of the usage of mechanical AI in the hospitality and tourism industries. The third section of the questionnaire recorded the responses to the anticipated perceptions of mechanical AI comparative to human employees and the customer experience. This section used a Likert scale ranging from 1 to 5 for recording responses. The number 1 was valued as strongly disagree and five as strongly agree. The mean score and the standard deviation for the responses were calculated to assess the aim of the study.

13.4 DATA ANALYSIS

The data recorded with the help of the questionnaire were analyzed using analytical tools such as Microsoft Excel and SPSS. The data was analyzed to meet the objectives of the study. In Table 13.1, the demographic profiles of the responding population are represented. It was observed that most of the female students between the ages of 21–25 have responded to the survey. The responding population further comprises undergraduate students and Doctorate scholars in the hospitality and tourism field. Also, almost every (97%) of the hospitality aspirants/professionals are well versed with the technological advancements using mobiles/laptops/tablets for day-today work.

In Table 13.2, the responses on the awareness about the common mechanical AI systems in hospitality and tourism are presented. It is observed that the maximum proportion of the responding hospitality aspirants is aware of the mechanical AI systems mentioned in the questionnaires. Among the categories asked in the questionnaire, the service conveyer belts were responded not known to the highest proportion (15) of respondents.

Table 13.3 represents the anticipated perception of the hospitality aspirants for using mechanical AI systems in the Hospitality and tourism industry. The responses collected on a Likert scale ranging from 1 to 5 as strongly disagree to strongly agree, respectively, denote that the aspirants of the Hospitality and Tourism industry strongly agree with the features of mechanical AI as Mechanical AI can be multi-lingual to the guests, followed by the statement that Mechanical AI is more efficient and faster

TABLE 13.1 The Demographics

Characteristics	Frequency	Percentage (%)
Gender		
Male	168	54.9
Female	230	75.1
Total	**398**	**130**
Age		
Below 21	47	15.3
21–25	129	42.1
26–30	114	37.2
Above 30	16	5.2
Total	**306**	**99.8**
Education Position		
School-going	33	10.7
Undergraduate	143	46.7
Post-graduate	22	7.1
Doctorate	108	35.2
Total	**306**	**99.7**
Residence		
Urban	164	53.5
Sub-Urban	124	40.5
Rural	18	5.8
Total	**306**	**99.8**
Do You Own a Personal Laptop/Mobile/Tablet		
Yes	297	97.0
No	9	2.9
Total	**306**	**99.9**

in calculations. Also, it is observed that the aspirants believe Mechanical AI makes fewer mistakes than Humans and are willing to be served with the mechanical AI systems.

Also, the responding population decides on the last statement in the questionnaire that with all the goodness in the AI mechanical systems, these cannot replace humans as these systems cannot understand human emotions and cannot be polite as per the situational responses.

TABLE 13.2 AI Awareness

Sl. No.	Mechanical AI Type	Percentage of Awareness of the Machine in F&B Service Industry	
		Yes Percentage (n = 306)	No Percentage (n = 306)
1.	Chatbots (dominos)	90.8	9.15
2.	Voice activated technologies (Siri, Alexa, Google Assist)	99.0	0.98
3.	Biometric systems (attendance/ identification)	100	0
4.	Hosting/welcoming machines	98.6	1.3
5.	Order taking machines/robots	99.3	0.65
6.	Service machines	97.7	2.28
7.	Intelligent vending machines	98.3	1.63
8.	Service conveyor belts	95.0	4.9
9.	Automatic bill settlement machines	97.7	2.28

TABLE 13.3 Anticipated Perceptions and Mechanical AI

Mechanical AI in Hotels	Mean	Standard Deviation
Mechanical AI can perform efficiently in complaint handling.	2.47	1.087
Personal information of the guests is safe with the application of AI.	3.33	0.993
Mechanical AI can serve to create a personalized hospitality experiences to the guests.	2.44	1.073
Mechanical AI is way faster than humans.	3.19	1.043
Mechanical AI makes less mistakes than humans.	3.49	0.949
Mechanical AI is more efficient and faster in calculations.	3.78	0.938
Mechanical AI can provide more accurate information of the regular guests.	3.1	1.153
Mechanical AI can be multi-lingual to the guests.	3.99	1
Mechanical AI can be more friendly than humans.	2.5	1.032
Mechanical AI can be more polite than humans.	1.27	0.378
Mechanical AI could be costlier in terms of capital investment and electricity/resources consumption.	3	0.926
Mechanical AI could have more chances of malfunctioning.	2.68	0.922
Mechanical AI can understand guest emotions.	0.3	1.247
I would love to be served by a mechanical AI.	3.14	1.13
Mechanical AI can totally replace humans in hospitality and tourism industry.	0.08	1.084

13.5 RESULTS AND DISCUSSIONS

The data analysis and the results show that the hospitality aspirants are technologically advanced and even well versed with the AI systems and the applications of AI systems in the hospitality and tourism industry. Also, it is observed that the young generations aspiring their career in the hospitality and tourism industry believe in the mechanical AI systems to cater for the guest in a better way and create a better hospitality experience in the various aspects of Hospitality and Tourism products. Also, it is concluded with the help of data received in the questionnaire that all the hospitality aspirants are at a common viewpoint all together that mechanical AI cannot replace human existence in the hospitality and tourism industry. Humans shall be required to oversee or control the AI systems installed at any point in the hospitality and tourism sector.

13.6 RECOMMENDATIONS OF THE STUDY

The study recommends that the study's results be taken as readiness for better management of human resources in the future. The aspirants of the hospitality and tourism industry today are the leaders/managers in the future. The trust shown by these aspirants in advanced technologies like AI systems and their applications in the various aspects of the hospitality and tourism industry shows that the future workforce is ready to adopt the technology into day-to-day operations.

The research results also advise that the university curriculum must include courses related to technological advancements to satisfy future job requirements in the hospitality and tourism service sector. The findings have significance and recommend future investigation. The study also recommends that the hospitality and tourism sector balance the activities of the human workforce and AI systems so that these systems can work on monotonous tasks with proficiency, whereas the human workforce keeps on delivering warm, personalized services to the guests.

KEYWORDS

- **aspirants and human workforce**
- **automation**
- **data analysis**
- **hospitality**
- **mechanical AI**
- **tourism**

REFERENCES

1. Qiu, H., Ye, B., Bai, B., & Wang, W., (2015). Do the roles of switching barriers on customer loyalty vary for different types of hotels? *International Journal of Hospitality Management, 46*, 89–98.
2. Albuquerque, D. D., Shukla, V. K., Verma, A., Tyagi, S. K., & Sharma, P., (2020). Enhancing sustainable customer dining experience through QR code and geo-fencing. In: *2020 International Conference on Computation, Automation and Knowledge Management (ICCAKM)* (Vol. 2020, pp. 190–196). doi: 10.1109/ICCAKM46823.2020.9051470.
3. Grobbelaar, W., Verma, A., & Shukla, V. K., (2021). Analyzing human-robotic interaction in the food industry. In: *Journal of Physics: Conference Series* (Vol. 1714, No. 1, p. 012032). IOP Publishing.
4. Naumov, N., (2019). The impact of robots, artificial intelligence, and service automation on service quality and service experience in hospitality. In: *Robots, Artificial Intelligence, and Service Automation in Travel, Tourism and Hospitality*. Emerald Publishing Limited.
5. Ashcroft, K., Barrow, F., Lee, R., & MacKinnon, K., (2012). Mindfulness groups for early psychosis: A qualitative study. *Psychology and Psychotherapy: Theory, Research and Practice, 85*(3), 327–334.
6. Punhani, R., Arora, V. P. S., Sabitha, A. S., & Shukla, V. K., (2021). Segmenting e-commerce customer through data mining techniques. In: *Journal of Physics: Conference Series* (Vol. 1714, No. 1, p. 012026). IOP Publishing.
7. Punhani, R., Arora, V. P. S., Sabitha, S., & Kumar, S. V., (2021). Application of clustering algorithm for effective customer segmentation in e-commerce. In: *2021 International Conference on Computational Intelligence and Knowledge Economy (ICCIKE) 2021* (pp. 149–154). doi: 10.1109/ICCIKE51210.2021.9410713.
8. Ivanov, S., Webster, C., & Berezina, K., (2017). Adoption of robots and service automation by tourism and hospitality companies. *Revista Turismo & Desenvolvimento [Journal of Tourism & Development], 27, 28*, 1501–1517.

9. Mohammed, U. S., Shukla, V. K., Sharma, R., & Verma, A., (2022). Next step to the future of restaurants through artificial intelligence and facial recognition. In: *Cyber Intelligence and Information Retrieval* (pp. 89–101). Springer, Singapore.
10. Shukla, V. K., & Verma, A., (2019). Enhancing LMS experience through AIML base and retrieval base chatbot using R language. In: *2019 International Conference on Automation, Computational and Technology Management (ICACTM), 2019* (pp. 561–567), doi: 10.1109/ICACTM.2019.8776684.
11. Cioffi, R., Travaglioni, M., Piscitelli, G., Petrillo, A., & De Felice, F., (2020). Artificial intelligence and machine learning applications in smart production: Progress, trends, and directions. *Sustainability, 12*(2), 492.
12. Misra, N. N., Dixit, Y., Al-Mallahi, A., Bhullar, M. S., Upadhyay, R., & Martynenko, A., (2020). IoT, big data and artificial intelligence in agriculture and food industry. *IEEE Internet of Things Journal*.
13. Upchurch, M., (2018). Robots and AI at work: The prospects for singularity. *New Technology, Work and Employment, 33*(3), 205–218.
14. Ivanov, S., & Webster, C., (2019). Robots in tourism: Are search agenda for tourism economics. *Tourism Economics* (in press). https://doi.org/10.1177/1354816619879583.
15. Del, R. J. S., Moctezuma, D., Conde, C., De Diego, I. M., & Cabello, E., (2016). Automated border control e-gates and facial recognition systems. *Computers & Security, 62*, 49–72.
16. Burgess, J., (1982). Perspectives on gift exchange and hospitable behavior. *International Journal of Hospitality Management, 1*(1), 49–57.
17. Lori, J. S., & Mark, R. T., (2018). From satisfied to memorable: An empirical study of service and experience dimensions on guest outcomes in the hospitality industry. *Journal of Hospitality Marketing & Management, 27*(2), 178–195. doi: 10.1080/19368623.2017.1306820.
18. Dogan, S., & Vatan, A., (2019). Hotel managers' thoughts towards new technologies and service robots' at hotels: A qualitative study in Turkey. In: Cobanoglu, C., Cavusoglu, M., & Corbaci, A., (eds.), *Advances in Global Business and Economics* (Vol. 2, pp. 382–399). Sarasota, FL: ANAHEI Publishing.
19. Chi, O. H., Gursoy, D., & Chi, C. G., (2022). Tourists' attitudes toward the use of artificially intelligent (AI) devices in tourism service delivery: Moderating role of service value seeking. *Journal of Travel Research, 61*(1), 170–185.
20. Tavitiyaman, P., Zhang, X., & Tsang, W. Y., (2022). How tourists perceive the usefulness of technology adoption in hotels: Interaction effect of past experience and education level. *Journal of China Tourism Research, 18*(1), 64–87.

CHAPTER 14

Digitalization and Customer Service Standards: A Service Industry Perspective

ROBIN SHARMA,[1] AMIT VERMA,[2] and VINOD KUMAR SHUKLA[3]

[1]*Department of Humanities, Arts and Applied Sciences Amity University, Dubai, UAE*

[2]*HTMi Switzerland Dubai, United Arab Emirates*

[3]*Department of Engineering and Architecture, Amity University, Dubai, UAE*

ABSTRACT

The chapter highlights the importance of customer-oriented industry standards and the impact digitalization is bringing to the industry. The service standards are considered one of the decisive factors for any service-related environment and are equally crucial for the guests who are staying or using the services. Total Quality Management and continuous improvement are the much-sought terms in the epitome of service and benchmarking of the standards. The industry has always been striving for a convenient and technology-based solution to improve efficiency and accuracy by raising the service quality and its benchmarks. Digitalization is a higher standard choice for service providers is juxtaposed to the actual customer experience and analyzing the accurate priority level for technology or customer service in the hospitality industry. The chapter also compares the technology-based solutions that are usually considered enhancing the service experience to customer service and overall customer satisfaction.

Artificial Intelligence for Smart Technology in the Hospitality and Tourism Industry.
Vinod Kumar Shukla, Amit Verma, and Jean Paolo G. Lacap (Eds.)

14.1 INTRODUCTION

The service industry is influenced and constantly evolving with the developing technological solutions. Starting from the early years of innovations, from electricity, hot water, telephone, television, and tablets, the service industry is widely accepted in terms of offering better, quicker, and accurate service matching to the brand standards. The industry seems fast-paced, actively adjusting to the guests' requirements and continuing its services from switchboards to smartphones, AI, IoT, and cloud-based network solutions. The economic aspect of technology, fast-paced life and trending digitalization has forced the masses to consider high-end service possible only with technology usage. The reservation has moved to the fingertips with the innovation of smartphones, relevant apps, internet, secure network, accuracy, record keeping, and the strategically planned and implemented choice to provide a high-end service experience to the guests. The choice of online booking, reservations, and check-in are some of the early results of implementing technology to fasten the waiting time and provide quality time to the guests. The industry professionals are eager to find opportunities in the guest's cycle that may provide convivence and remove the idle time by providing several options of entertainment and services, which may result in higher spending from tourists.

14.2 DIGITALIZATION IN SERVICE INDUSTRY

There was a time when the only way to find out about a hotel was to travel and experience it first-hand. In today's digitalized world, a customer can surf the internet and understand the peculiarities of hotel services, rating, and grading, with the latest reviews from the real customer. His own generated content is easily accessible. The days had gone by when hotel booking was made either by contacting the hotel directly, walking in, or seeking travel agent support. The internet has moved them all to the online travel agency, self-check-in, customer portal, and applications.

As per digital adoption worldwide, 62% of the world population uses the Internet [2]. The increase in the internet user population is considered to be directly linked to the country's economic condition and the hospitality industry, as depicted in Figure 14.1. The tourism destination countries and origin countries are equally impacted by tourism between the countries leading the hospitality industry's economic and technological influences and advancements. The Internet has revolutionized the world and brought many significant developments for guests and service providers. The closely-knit network of the Internet has seamlessly connected users from all over the world in a web of information and empowerment. It has changed the way customer interacts with the hotels and vice versa. Social media has derived multi-fold benefits from the apps such as Instagram, Twitter, and Facebook by directing the brand launch, identity, and promotions to feedback and customer communication. Recently, the bot services on the messaging apps have also started ruling the communication support and proving beneficial.

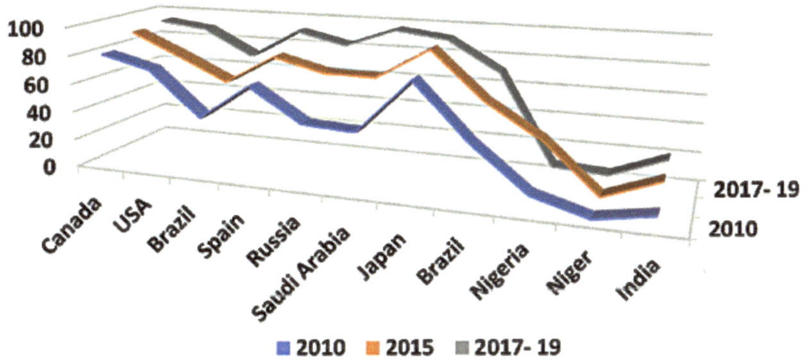

FIGURE 14.1 Internet usage chart worldwide [1].

14.2.1 IMPERFECTIONS OF INNOVATIONS IN THE SERVICE INDUSTRY

The hospitality industry is known for its intangible experience generated through human interaction and making guests stay pleasant. Technological advancements have raised the question of compromising the personal touch in the hospitality industry. Below are a few ineptitudes in implementing innovation in the service industry:

1. **Undefined Expectations:** Each guest is different, and on their next trip, they have a different set of expectations. The changing expectation would only be adding more options to the program to prove its vitality. However, the undefined expectation is the

one that the guest himself/herself does not know unless needed. In such circumstances, human support is needed.

2. **Fatal Errors and Delays:** The adaptation of technology has been successful with its continued maintenance and dependency on the set of support services relatively low but has a possibility of fatal error and can hold the system.
3. **Data Security Issues:** Data security and theft are some of the concerns of any service provider where the number of guests their details and collectively in a hotel the database is increasing in hundreds every day.
4. **Absence of Cultural Exchange on Guests' Level:** There is a gap in the experience level of the guest and machine which cannot participate in the cultural exchange with the guest.
5. **Shifting Focus from Service to Digitalization:** Along with the trending technology around the world, the hotel industry has peer pressure to upgrade the innovative style of services even sometimes, if it is not the priority; therefore, we may see several sets of gimmicks to lure the customer only.
6. **Standards, Innovation, and Economic Mismatch:** The service standards dwell on the expectation of the guests and an innovative check-in system or Alexa support may turn out financially expensive for small hotels. It may be difficult for the guests of old age or non-technology literate who might struggle to handle the system.
7. **Frequent Changes:** Force staff to upgrade and old disbarment devices, technology, and processes.

The continuous adaptation may never let the processes be defined and seek a focused approach toward the upcoming technology, which may be better and accepted but would make the organization move to the new set of processes and systems. This amount of energy and effort should be intentionally focused on raising the guest's experience, not competing with the trending industry gimmicks.

This does not mean that the technology is not supported. Nevertheless, it insists that the financial viability, country priorities in tourist regions, and the return on investment are based on overall service standards based on real customer experience compared to the non-digital yet highest standard of service.

The adaptation of technological features and its impressive advantages are not neglected in service standards. However, the customers who do not want or could not use the technology are left for double standards unless foolproof technology support is not available to avoid the bad experiences due to innovative approaches. For example, the user profile and preferences are one of the sought-after details of the hotels to analyze and trace the client's choices. The chances of finding such analysis inaccurate are due to the dependency on user-provided data. The profile can be used by several house members or may have no relevant user.

Furthermore, the world is still not accurate on the web network and always deploys its technology and manual support operations. This is a common feature seen for the self-check-in kiosk operation. You may find a human assistant to guide a set of populations (old/young age technology literate/illiterate, language issues and disabled) along with the system errors and network issues, etc. This has further geared the innovators, and AI has started loading the machines with sensors and learning. The intelligent device is intelligent to hear and act how your preferences are analyzed but may easily commit a mistake in identification.

14.3 IMPROVING GUEST EXPERIENCE BY INTRODUCING ARTIFICIAL INTELLIGENCE (AI)

The digital world is creating new standards in customer service by introducing AI. Companies use smart technologies to interact with their customers in the hospitality industry, where the human touch is inevitable. The introduction of chatbots, voice-activated assistants, luggage delivery robots are a few of the many introductions in the industry. The Future Hotel is evidence of dominated human interactions and experience through technology in the hospitality industry. The famous voice-based assistants are enforcing their implantations in all devices. Amazon Echo and Alexa have changed the dynamics of digital customer service by using AI. Technology adaptation has to lead a robot to handle the hotel completely, which sometimes raises the test of the misconception of AI ruling the industry [3]. The hospitality industry to adopt technology shows that they are committed to operational excellence and are ready to transform their daily operations. When a brand plans to introduce AI in its operation, they need to have a transparent, thought-out process on what they want to achieve. The goal should be to enhance the customer experience. On the other hand, customers should also understand that they are interacting with an AI-enabled system to be aware of the limitations. For example, the robots may only convey the set of algorithms fed by a human for the response of fun activities by the human who is still the point of choosing the activities. Customers enjoy interacting with technology as it brings excitement and quick response (QR) [4].

The customer today is more demanding, detailed than it was before. The industry needs to find new ways to keep customers happy and engaged. Hotels introduced a first its kind "Robot Concierge" Connie, who can recognize and respond to questions that guests might ask of a concierge. Each customer interaction allows it to learn, adapt, and improve recommendations. Robots greet guests and tell them about hotel amenities, services, and operations hours. AI tools can also be implemented to interact with customers.

Moreover, IoT, Blockchain, and other technological solutions colloquially support the operations. IoT trend in enterprises has also enhanced, as depicted in Figure 14.2 in 2020 [5]. The change has come as the customer demands fast, reliable, and responsive service. We should not rule out that the customer still requires an experience-driven service wherein core human values of empathy and trust exist. Therefore, organizations need to balance the Bot and humans touchpoint. The services of AI are throbbing in 2020 yet are added to supplant the services, not to remove the human touch [6].

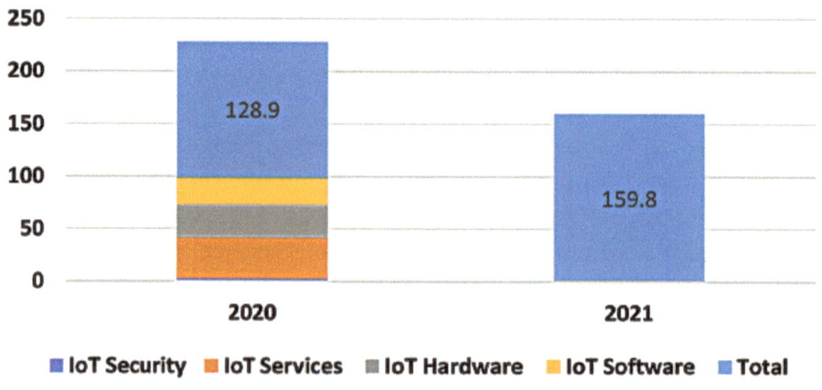

FIGURE 14.2 IoT enterprise spending 2020–2021 [5].

The customers connect with the brand in various ways. The most preferred method now is digital support, wherein they can notify about the request or problem rather than connecting through traditional channels. With the expansive growth of AI, bots may revert immediately to your query if it is already identified programmed with limited options. Several aspects can be developed through the help of AI in the Hospitality Industry, and only a few have dedicated options to follow. However, the rest do not, which needs a diligent approach. The use of AI will demonstrate customer behavior regarding purchasing, travel preference, patterns, location, preferred hotel rating enquiries, payment options.

14.4 TECHNOLOGICAL TRENDS THAT CAN ELEVATE GUEST EXPERIENCE IN HOSPITALITY INDUSTRY

The intelligent automated data collection of consumers will allow organizations to understand the needs and expectations of their audience and will enable them to customize guest experiences. Customers like to be in control when they arrive at hotels. The use of mobile applications gives them the sense of being in control. The hotels nowadays are adopting and providing digital solutions to the guest. Hotel companies provide most of their services through mobile applications like a smart key, wake up call, ordering food, etc. The hotels are introducing Robots in their service delivery approach. Robots store guest luggage, deliver guest food orders and cut down on human capital to do monotonous tasks, improving employee productivity.

14.4.1 SMART KEY

Going plastic-free, enabling guests to check-in and access their rooms using their mobile phones (NFC technology) [7]. This will give the guest freedom to roam freely in the Hotel. The Hotel, on the other hand, can benefit from this technology by saving one guest room key card and will be able to collect guest data.

14.4.2 SELF-SERVICE

The Self-Service kiosks enable guests to order food through Mobile apps or iPads placed in guest rooms to select the services provided to be booked through the help of self-check-in kiosks. Sometimes it is a hassle also for a few customers to operate a TV and book the services throughout without getting any support [8].

14.4.3 MOBILE HUBSPOT IN THE GUEST ROOM

Allowing guests to connect the phone to operate the TV, blinds, and sound system. Guest likes to be in control when staying in hotels. The self-service food kiosk and applications will increase the accuracy of guest orders and reduce the time spent on each phone call by the order taker. By providing smart guest solutions in the guest room, the amount of money spent on entertainment channels can be reduced as the guest will be able to stream anything from their phone or laptop [9].

14.4.4 GUEST APPLICATIONS

A significant portion of the technology is exposed to the guests through the apps and kiosks operating apps. An essential part of guest involvement with technological advancement is selecting the options and getting a notification to loyalty programs, welcome guides, and other services for guests [10].

14.5 RISK ASSOCIATED WITH DIGITALIZATION

The Hospitality industry is a prime target for cyber breaches and data privacy risks [11]. The vital information which is collected by each operating segment of the Hospitality Industry is personal guest data. With technology and digitalization, hotels are at risk of cyber-attacks. One of the top-notch Hotels announced a cyber-attack wherein 25 million guest passport numbers and 380 million guest unique, personal information was compromised. Likewise, the number of credit card transactions in hotels are colossal, which tempts cybercriminals to breach the interface. The cyber fraud attack in South Africa in 2012–2013 was focused on Point-of-Sale Systems wherein the attacker used "Dexter" malware to compromise security levels [12]. Thousands of shops, hotels, and restaurants fell victim to it. The malware skimmed and transmitted credit card magnetic strip information, allowing duplicate cards to be used for fraudulent transactions.

14.5.1 NEGATING RISK FACTOR

The adapted technologies come with their protective mechanism and get compromised after a stage where either the loophole is discovered by the attacker first results in the compromised situation else translated as the security update in the software, but the risk remains unknown which a hotel can reduce the risk of cyber-attacks by avoiding risk from phishing attacks, Dark hotel hacking, malware, ransomware, denial of services or DDOS attacks, Point of sale and payment card [13]. This is usually managed with the secured log in channels, the latest version of antivirus software, educating staff and customers on how to avoid cyber-attacks, risk assessment, etc.

14.5.2 COVID-19 CRISIS ON DIGITALIZATION

The COVID-19 has forced industries to undergo a severe fallback due to high investments and minimum/zero returns of the companies. The world has taken a halt with the travel restrictions, and hotels suffered badly, many got closed forever, and few hotels kept bleeding for years awaiting their recoveries. In such a turmoil, technology took the front seat to drive the change and support the operations effectively in the part of the world where the kiosks were just a gimmick and always needed human support.

WTTC (World Travel and Tourism Council) stated the critical digital solutions to restore international mobility [14]. Technology revived the hotel industry in the COVID times by fastening the reopening through quick adaptation by the government and supporting industries. This has developed trust among the tourists. In the COVID time, approximately 75% of customers prefer to purchase again from a company that knows their name and preferences [15]. This has allowed many companies to adopt technological growth and invest in the technology, and the experts suggested CEOs move to cloud technology [15] quickly.

14.6 CONCLUSION

Digital transformation of customer service is a trend that will impact all the subsidiaries and companies to improve the efficient support system. The digital-first strategy will provide more channels for customers to interact. To have a competitive advantage, the Hospitality Industry must shift to technological advancements. The adaptation must be cognitive in terms of real utilization and support of the client not only the hoarding of the adaptation of technology. The trends discussed in the chapter are more applicable to large scale hotels. Hotels catering to the specific demands of a niche client, such as natural escapes and machan hideouts in the wildlife safari, etc., might not require immediate and significant shifts in technology. However, at the same time, they should not ignore the technological advancements that could be a support mechanism in the operations.

KEYWORDS

- **COVID-19**
- **customer service**
- **digitalization**
- **hospitality**
- **satisfaction**
- **self-service**
- **technology**

REFERENCES

1. Ritchie, H., & Roser, M., (2017). *Technology Adoption.* Our World Data. [Online]. Available: https://ourworldindata.org/technology-adoption (accessed on 07 September 2023).
2. *Digital Around the World.* DataReportal – Global Digital Insights. https://datareportal.com/global-digital-overview (accessed on 07 September 2023).
3. *AI in Hospitality in China: Are Robots Taking Over the Hotel Industry*, (2019). *Daxue Consulting – Market Research China.* https://daxueconsulting.com/ai-hospitality-china/ (accessed on 07 September 2023).
4. *Could Robots Become the Next Travel Retailers?* https://amadeus.com/en/insights/blog/robots-become-next-travel-retailers (accessed on 07 September 2023).
5. *Global IoT Spending to Grow 24% in 2021, Led by Investments in IoT Software and IoT Security,* (2021). IoT Analytics. https://iot-analytics.com/2021-global-iot-spending-grow-24-percent/ (accessed on 07 September 2023).

6. Bob, R., & Sarah, A. *Hotel Robotics Thriving in 2020* | Hospitality Net. https://www.hospitalitynet.org/opinion/4100756.html (accessed on 07 September 2023).
7. *6 Ways Hotels Can Benefit From NFC Technology*, (2018). Infraspeak Blog. https://blog.infraspeak.com/6-ways-hotels-can-benefit-nfc-technology/ (accessed on 07 September 2023).
8. *Contactless Tech in Hospitality 2020*. https://research.skift.com/report/contactless-tech-in-hospitality-2020/ (accessed on 07 September 2023).
9. *Smart Energy Systems Can Help Ensure Guest Satisfaction* | Hotel Management. https://www.hotelmanagement.net/tech/why-smart-energy-systems-ensure-guest-satisfaction (accessed on 07 September 2023).
10. Ryan, K. *Self-Service Sells: The Rise of Self-Service in the Hotel Industry | Hospitality Net*. https://www.hospitalitynet.org/opinion/4093051.html (accessed on 07 September 2023).
11. PricewaterhouseCoopers, *Cyber Risk: Hotels*. PwC. https://www.pwc.com/m1/en/publications/documents/cyber-and-data-security-in-the-hotel-industry.pdf (accessed on 07 September 2023).
12. *Dexter Payment Card Malware Strikes South Africa*, (2013). BBC News. [Online]. Available: https://www.bbc.com/news/technology-24550505 (accessed on 07 September 2023).
13. *Hotel Cybersecurity: A Guide to Avoiding Threats* | Cvent Blog. https://www.cvent.com/uk/blog/hospitality/hotel-cybersecurity (accessed on 07 September 2023).
14. News Article | World Travel & Tourism Council (WTTC). https://wttc.org/News-Article/WTTC-identifies-digital-solutions-for-governments-worldwide-to-significantly-restore-international-mobility (accessed on 07 September 2023).
15. Baig, A., Hall, B., Jenkins, P., Lamarre, E., & McCarthy, B. (2020). The COVID-19 recovery will be digital: A plan for the first 90 days. *McKinsey Digital, 14*. [Online available on: https://www.mckinsey.com/capabilities/mckinsey-digital/our-insights/the-covid-19-recovery-will-be-digital-a-plan-for-the-first-90-days] (accessed on 07 September 2023).

CHAPTER 15

Artificial Intelligence and Individual Privacy

GAGANDEEP KAUR,[1] PRASHANT CHAUHAN,[2] and
VINOD KUMAR SHUKLA[3]

[1]*School of Law, University of Petroleum and Energy Studies, Dehradun, Uttarakhand, India*

[2]*Research Scholar, University of Petroleum and Energy Studies, Dehradun, Uttarakhand, India*

[3]*Department of Engineering and Architecture, Amity University, Dubai, UAE*

ABSTRACT

Civilization has witnessed constant deliberations on human rights. The development of technology followed the journey of human rights. Throughout human history, technology has been vital in the development of humans. In the age of technology, most aspects of humans have been impacted. However, certain fundamental rights have become vulnerable; privacy is one such right. Among the most cherished rights is the right to privacy, a fundamental component of the right to life and personal liberty. Modern technology allows us to live in a virtual world that we have created. Through this virtual world, we are connected and vulnerable to an invasion of our privacy. Artificial intelligence (AI) is another hybrid version of the technology that poses a more significant threat to the privacy of individuals. Using AI to create and manipulate personal data is a threat to the user's privacy in this world, where information is generated at every

Artificial Intelligence for Smart Technology in the Hospitality and Tourism Industry.
Vinod Kumar Shukla, Amit Verma, and Jean Paolo G. Lacap (Eds.)

level since it has no ethical constraints and may act autonomously. This chapter intends to examine the impact of AI and its challenges towards the right to privacy.

15.1 INTRODUCTION

Technology has played a pivotal role in shaping civilizational development. In current times, the individual is surrounded by technology so that human existence without technology is unimaginable. As humans have advanced, comparably, the innovation has seen fast movement. A few ideas which were additionally subject of exploration have come to fruition in functional life. As intelligence is the one that makes humans superior to other creatures of this world, it is only due to human intelligence that civilizations have been developed and progressed. With the use of its intelligence, humans have created several artificial things. Similarly, attempts have been made to create intelligence artificially, which is at par with human intelligence.

One such concept is artificial intelligence (AI). A concept that began to emerge in the mid-1950s has nearly touched upon every aspect of human life. AI is a hybrid of the fundamental mechanisms of computer programming. Several electronic devices have been incorporated with a specific level of AI in some of the other manners. As individuals live in a virtual world, it has exposed them beyond certain limits. Human rights are cherished as the fundamental Right of human beings. They have become vulnerable to being breached. As several decades of struggle have taken place to establish and protect human rights, today, individuals, whether knowingly or unknowingly, have crossed the limits allowing the breach of these rights. One such Right of prominent importance is the Right to privacy. Privacy, the subject of an individual's inner core, has virtually been exposed to possible attacks.

As AI uses data for processing information, the question is that when AI is allowed to access and use that individual's data, especially when AI is infused with automation, to what extent such data is secured from possible misuse and manipulation. Considering this factor, whether the use of AI should be prohibited or can a mechanism be developed to secure the individual's interest and strike a balance between individual rights and technology. AI raises the power and speed of analysis of personal information, which is a potential threat to privacy interests over time. Where AI advances, it amplifies the means to intercept and utilize the individual's information in such a manner that can foreclose the interests of private nature.

AI differs from ordinary software or an App, designed to perform particular functions. The functions are programmed by the creator and can be modified later. Together, these applications are bound to perform within the periphery they are designed to operate. The working mechanism of AI differs from the ordinary software or an App, despite the fact they form their basis on the fundamental of computer programming. AI works based on the algorithm. These can be designed to make the machine act independently; here, the problem arises: while performing functions, the AI will collect, analyze, and individual process data, putting individual privacy at risk.

Facial recognition systems outline privacy concerns. Recognition of facial traits by a machine has made rapid progress from blurred images of deer to rapid (though lacking in perfection) recognition of individuals. This has also contributed to a huge repository of data of photographs in digital format available through online social platforms, websites, licensed registries of device drivers, and vigilance cameras, among others. Numerous cities and airports in America have begun to implement facial recognition technology. China, however, has been criticized for implementing facial recognition with absolute control in Xinjiang and elsewhere, and calls have been made to ban its use. Many communities have banned facial recognition because of concerns over security and privacy. Many cities in California and Massachusetts, including Somerville, Northampton, Cambridge, and Brookline, have banned it. The application of facial recognition in bodycam used by law enforcement has been banned through legislative mandates in Oregon, New Hampshire, and California [1].

15.2 RISK INVOLVED IN ARTIFICIAL INTELLIGENCE (AI)

There is a significant risk involved in the application of AI. It can be understood based on the following points:

1. Firstly, the processing of information by a machine is no match as far as human intelligence is concerned, resulting in endangering the personal data of a larger public, making it a more significant

threat. In this virtual world, as most individual information is being collected in digital form, it is prone to possible hacking. The hackers can also manipulate the system for their ulterior motives.

2. Second is the level of accuracy by machine in determining the possible outcome. As the algorithms constitute the fundamental working mechanism of AI, the level and extent of accuracy are dependent upon its programming structure. As the machine becomes automotive in its action, the activities conducted by the machine cannot be intercepted by human intervention.

It's also worth noting that some of the potential harms from incomplete facial recognition is due to facial recognition technologies. Facial recognition also helps to find similarities to the other demographic groups, such as people with darker skin color, Asian origin, non-binary people, and old age. Concerning law enforcement investigations, due process is eroding and having chilling effects upon protests. The erosion of practical obscurity regarding facial recognition is harm that is broadly shared by the public. The use of facial recognition not only invades the private personal space but also creates space for discrimination on a demographic and racial basis, resulting in further violation of the fundamental principles of Human Rights.

An algorithm may provide a range of probable matches depending on how the program is designed. Alternatively, based on the matching algorithm, the algorithm may assess the plausibility of identifying the person correctly. Several systems, like Amazon Recognition, let the end-user assess a level of confidence that the image is a match. However, the company has been non-transparent about the metrics that are applied. An image's quality plays a significant role in how well the algorithm identifies the face and identifies to whom it belongs. Low-resolution images, poor lighting conditions, and twisted or obscured faces are all factors that result in lower results [2]. Although facial recognition is based on AI, the outcome can depend upon several extraneous factors or circumstances. These factors can produce a distorted result, far from the actual scenario. Hence, the real problem lies not in the technological aspect but rather the application part. The application part can create a threat-like situation for the Individual's privacy, ultimately posing a challenge to the very core of Human Rights.

15.3 ARTIFICIAL INTELLIGENCE (AI): THE CONCEPTUAL FRAMEWORK

The formation process of AI can be understood by the representation in Figure 15.1.

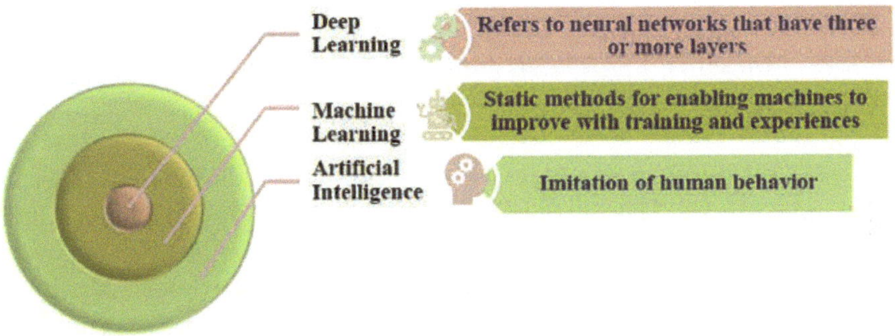

FIGURE 15.1 Graphical representation of artificial intelligence.

The components, as depicted in Figure 15.1, have been discussed briefly under the subsections.

15.3.1 DEEP LEARNING (IMITATING HUMAN BRAIN)

Deep learning acts as the bridge between AI and machine learning. It acts as a window for utilizing the data by machine for self-decision making. "Artificial Intelligence: A Modern Approach" by Peter Norvig, Google's Director of Research, is a textbook acclaimed for its coverage on AI. The talk he gave in the year 2016 was under the heading "Deep Learning and Understandability versus Software Engineering and Verification," which explained deep learning like Yoshua's, looking at the perspective for abstraction accredited by deeper networks. This is a type of learning where you form representations with more than one level of abstraction instead of getting an undeviating response from the input.

A technical report, "Learning deep architectures for AI," in the year 2009, discussed deep learning in more detail where he emphasizes the importance of the hierarchical structure in determining features. Deep learning methods can learn feature hierarchies by composing low-level features from higher-level features. Learning multiple levels of abstraction automatically allows systems to map inputs to outputs directly from data, rather than relying on manually crafted features. There is a clear indication that although deep learning is created through human intervention, it does not rely on the process that created it. It can understand the complex functions and act accordingly.

15.3.2 MACHINE LEARNING (ACQUIRING KNOWLEDGE THROUGH EXPERIENCE)

We know that modern machines work based on the combination of hardware and software. Hardware is the outer structure of the machine, which performs the specific functions for which the machine has been created. However, the machine cannot perform any function just mechanically. It requires programming. Programming means a set of instructions that will make the machine perform the function as per the desired functions to be taken from the machine. Hence, when we talk about AI, machine learning is based on the algorithm designed to process the collected data for the purpose it is to be used. As a branch of AI, machine learning develops applications that can learn from data and become more accurate over time without being programmed. Statistical algorithms are the steps in the data science process. By learning from massive amounts of data, algorithms can find patterns and aspects to make predictions and decisions. As more data is processed, the better the algorithm will be in making decisions and predictions.

15.3.3 MACHINE LEARNING

Machine learning involves the following steps in its functioning:

1. **Constructing and Adopting Set of Data:** In general, the training data is selected in such a manner to facilitate the machine to act with a problem-solving mechanism.
2. **Selecting Algorithm for Training Purpose:** The algorithm selection is based on the use of the machine for what kind of problem resolving is required.

3. **Adjusting the Algorithm for a Particular Model:** It is meant to attain more accurate results based on variables.
4. **Testing and Improvising the Model:** Lastly, to test the model's functioning created and by continuous observation, figure out the shortcomings and fill them.

15.3.4 ARTIFICIAL INTELLIGENCE (EMBODIMENT OF HUMAN BEHAVIOR)

Machine learning clubbed with profound learning results in AI. Prof. John McCarthy, founder of the AI laboratory at Sandford University, California, coined the term AI for the first time. Machines that perform tasks generally performed by humans are considered to be using AI. As a result, machines can learn through learning experiences and develop the ability to perform without human intervention. An artificial neural network, an algorithm inspired by the natural brain, learns from massive data sets using deep learning.

A deep learning algorithm performs a task repeatedly in the same way we learn from experience, tweaking the outcome each time to improve it. As a result of the many layers in neural networks that enable learning, we refer to the technique as "deep learning." Deep learning can be trained to solve any problem that requires "thoughts" to solve. AI is based on the framework of the human brain, which contains neural networks that help the human brain in learning by way of perception and experiences. Meaning thereby AI can be determined as making a data processing machine act and decide through a self-learning process. This process is based on the process followed by the human brain.

15.3.5 PATRICK WINSTON

Professor at the Massachusetts Institute of Technology (MIT), explained AI as using impediments to enable algorithms and exposing them by representations, linking thinking, perception, and action into loops. Where Mr. Jeremy Achin, CEO of DataRobot, stated that AI is the capability of a computer to perform tasks ordinarily performed by humans. AI systems rely on machine learning, deep learning, and rules to learn. Hence, based on the above opinions on AI, it can be concluded that AI is the outcome of Machine learning infused with Deep Learning. Only understanding both concepts will pave the way for understanding the working mechanism of AI. AI has now reached a level equal to human intelligence when machine learning and deep learning are combined. When AI becomes equivalent to Human Intelligence, it might pose a serious challenge to certain rights of Human beings.

15.4 RIGHT TO PRIVACY – LARGELY DISCUSSED; LEAST EXPLAINED

Nearly all countries and international organizations recognize that the right is an integral part of each individual's right to life and personal liberty. In the initial phase, the right to privacy was considered part of the right to life and personal liberty. With time, it has emerged as an independent right, taking center stage. The judiciary had a long journey in recognizing this right in its current form.

The most essential and cherished right is the right of life, which is *sine-quo-non* of the very existence of Human beings. At the same time, the right of individual liberty is considered to be embodied within the right to life. Over time, many facets of the right of life have emerged. One of the most important has been the right to privacy. Figure 15.2 depicts the evolution of the right of privacy through the Judicial approach. The Judiciary of the United States of America can be the torchbearer for the right of privacy. In the landmark of Wolf vs. Colorado [3], the court highlighted the constitutional protection accorded to the right of privacy. Whereas in India, in the case of Kharak Singh vs State of UP [4], Subba Rao, J. being the minority judge, acknowledged the privacy right within the ambit of the fundamental right. In the case of Govind vs. State of M.P. [5], the Supreme Court affirmed the right of privacy as a fundamental right. Whereas in the case of R. Rajagopal vs. Union of India [6], the Supreme Court more precisely brought the privacy right within the ambit of personal liberty. In People's Union for Civil Liberties (PUCL) vs. Union of India [7], the Supreme Court broadened the ambit of privacy and made it applicable to communications between individuals. In the case of District Registrar and Collector vs. Canara Bank [8], the Supreme Court stated that personal liberty, freedom of expression, and freedom of movement all facilitate and complement the existence of the right of privacy. Selvi v State of Karnataka [9] distinction was drawn between the menta and physical privacy. The most critical case in the establishment of the right of privacy has been the case of Justice K. S. Puttaswamy (Retd.) and Anr vs. Union of India and Ors [10]. The supreme court of India firmly stated that privacy is guaranteed by Article 21 of the Indian constitution. Hence, the Right to Privacy is a right of paramount importance with the right to life and personal liberty.

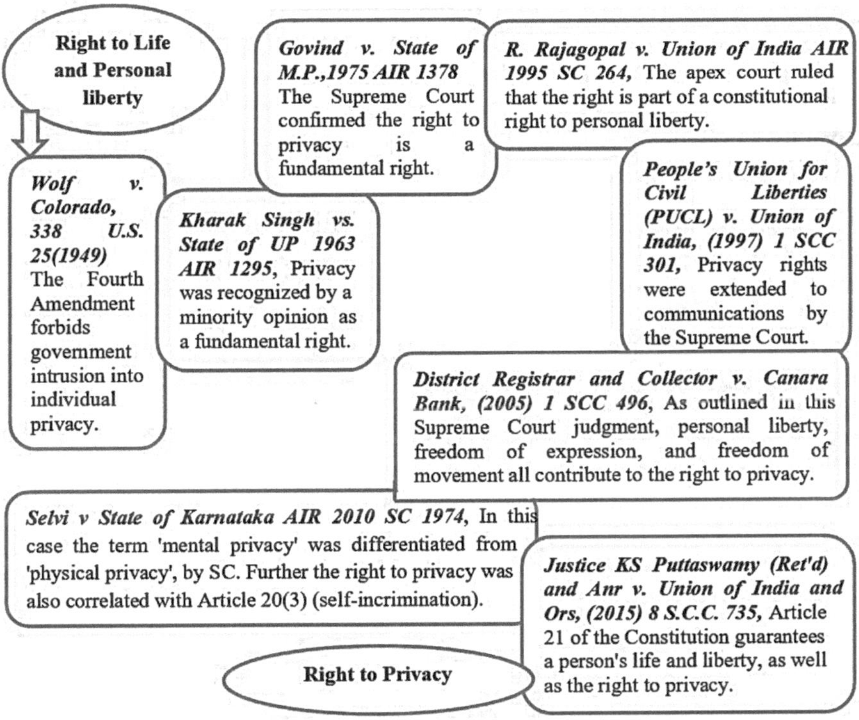

FIGURE 15.2 Chronological evolution of right of privacy: Landmark case laws.

15.5 ARTIFICIAL INTELLIGENCE (AI) VIS-A-VIS RIGHT TO PRIVACY: AN ANALYSIS

The matter of concern is that most of the data nowadays is stored on virtual platforms, i.e., the cloud. The collection, processing, and usage of such data are being performed using AI. It is good to the extent that AI is used to collect, process, and conserve data. However, when AI is automated, it may give rise to certain serious apprehension. Mr. Giovanni Buttarelli raised a very striking point. He said that "AI raises fundamental questions of accountability for outcomes, as well as accountability for collection and use of massive quantities of personal data" [11].

It is a valid point as AI works on the Algorithm. How will the machine be made accountable, or the person who constructed the Algorithm will be held responsible? AI somehow lacks the ethical aspect, and the manner of dealing with the issues by a machine and Humans is different. In computer science, machine learning algorithms are trained using an anonymization technique known as one-way hashing. Hashing is a process that converts data into a number to prevent the original data from being derived from it alone. For example, when the hash operation is performed on data records with the name "Robert" associated with them, the name "Robert" is turned into a numerical representation that is mathematically tricky or impossible to convert into the individual's name. There are many techniques used to anonymize, but none are foolproof. Data protection authorities in Europe have published detailed guidance on what can and cannot be anonymized using hashes.

Another drawback is the interference with the democratic process. Especially the case of Cambridge Analytica showed that the principal factor was "erosion of confidence in government and public institutions. Interference by Russia and voter manipulation by Cambridge Analytica in the 2016 presidential election played a large part in that public disaffection." [12] Mr. Julian Knight MP stated that by examining the processes that SCL used, it became apparent that most of them were well-established processes employing standard technologies. By downloading from public open-source libraries for data science, such as Scikit-Data visualization, forecasting, and analysis algorithms are well-established and widely used – SCL demonstrated a clear understanding of data science. SCL's data science activities comprised a majority of third-party libraries that the ICO observed. To determine which machine learning models, activation functions, and optimizers predicted any dataset with the highest accuracy, SCL used these libraries to investigate a variety of possible architectures, activation functions, and optimizers. The data science community is well versed in this procedure, and we do not believe it contains any processes or technologies exclusive to SCL [13].

These above-stated processes used a variety of Algorithms to analyze the appropriate US voter GSR data in large amounts and put together "predicted" scores based on factors such as partisanship and other criteria to make applicable to all individuals in the database [13]. This shall have a disastrous effect on democracy, and the democratic process itself will be eroded, as well as the privacy of the individual will also be endangered. The individual's data is open to manipulation, affecting their right to privacy and their political and social rights.

An expert seminar was conducted on AI and the right to privacy by the UN Human Rights High Commissioner. The National Human Rights Commission of India states that AI technologies can offer enormous social benefits. However, they also pose serious human rights challenges, notably concerning data mining techniques. The Chair recalled the judgment of the Supreme Court of India of 2017 recognizing the right to privacy as being protected by Indian constitutional law. It noted the need to establish a framework for data protection, such as sector-specific rules, to regulate AI technologies and encourage developers to follow international privacy standards [15]. The reason for apprehension is that there is a vast difference between human intelligence and AI. The mechanical aspect of AI is far superior to human intelligence in processing data is concerned. Although if we consider the interactive aspect, human intelligence is far superior to emotional and ethical aspects.

15.6 CHALLENGES POSED BY AI TOWARDS RIGHT TO PRIVACY

Although there are numerous challenges that might pose a threat to the right to privacy, some of the major issues are as follows:

1. **Artificial Intelligence (AI):** As AI is based on algorithms, the transparent algorithm's programming structure is not available. It is only the outcome. Hence, there is a lack of transparency concerning programming. The primary reason for keeping the algorithm a secret is in the creator's interest, be it technical, economical, or Intellectual. Comprehensibility and accessibility of information are the primary components of transparency. Many algorithms do not readily disclose their internal logic. The purpose of proprietary algorithms is to provide a competitive advantage or protect national security or privacy. In this sense, transparency can

conflict with various ethical issues, such as protecting the privacy of individual's data and preserving the independence of organizations [16].

2. **Cybersecurity Threats:** Technology specifically about the internet has transformed drastically, facilitating the dissemination of information and leaving the internet or cyberspace vulnerable to possible attacks. Hackers are always looking for opportunities to break into cyberspace, endangering digitally stored data. AI is a branch of technology that is open to being controlled and manipulated, and if it is used in cyberspace, it can pose a serious threat. Based on this data diet vulnerability, new attack vectors are opened up by incorporating AI into national security surveillance or cyber security. AI-based surveillance systems can be fed disinformation by adversaries, creating unwitting automatic double agents [17].
3. **Bias, Injustice, and Discrimination:** As the action of AI is dependent on various extraneous factors. There is an excellent chance of error in its functions. Specifically with a particular task like facial recognition, which gives rise to the fact that the outcome of AI might result in bias and discrimination based upon biological or geographical factors, which shall be against the fundamental principles of Human Rights. Data analysis can improve services, advance research, and combat discrimination. Nevertheless, such analyses may also raise valid questions regarding those vulnerable to harm or who are treated differently. If automated decisions affect people's eligibility for housing, employment, and other core services, they can be amplified when used sensitive data (such as race, gender, or familial status) [18].
4. **Issues Concerning Legal Personhood:** Several voices have been raised to bring AI within the parameters of personhood or, rather, to be more precise, legal personhood. AI cannot be considered at par with the natural person being a creation of a Natural person itself. Being independent in its action will not be represented by a natural person. In both, the situation granting legal personhood is a tricky task. Therefore, human rights will be weakened when compared to synthetic persons. The next step is to examine whether there are moral benefits, such as achieving moral objectives, that offset

these risks or costs. The author believes that AI does not have a moral obligation to be recognized as legal. The extension of legal personhood to robots is not recommended because it would incur high costs and have few moral benefits [19].

5. **The Issue of Intellectual Property:** In terms of AI, intellectual property is crucial. AI emulates human intelligence but is also created through the application of the human brain. The question arises as to who shall hold the IP rights if an invention is created through AI. The AI, or the person who created the AI. Because AI has acted independently, AI has many intellectual property issues. For example, who owns the works or inventions created by AI? Could AI's inventions be considered prior art? If AI must gain knowledge from a dataset, who owns it? Who should be liable in the case of creative and innovative projects developed by AI infringing upon other people's rights or violating other legal provisions? [14].

15.7 CONCLUSION

There is no question that technology cannot be undermined in the current time, especially the technology concerning AI. The use of technology is inevitable, but the apprehensions raised have to be redressed. The use of AI has increased on the technical level, but attention must be drawn on as to what kind of drastic impact it can cause on individuals' fundamental rights, which had taken decades to evolve and establish. Although there are specific regulatory mechanisms as developed by the European Union efforts have been made to enhance and strengthen the regulatory mechanism. However, these guidelines are lacking in their basic framework. Goal-oriented specific laws or guidelines are necessary to safeguard individuals' privacy. Even the landmark judgment of Justice KS Puttaswamy (Retd), which categorically establishes the right to privacy as a peculiar right, somehow has fallen short of considering the threat which is bound to arise with the rapid intrusion of AI into Human lives. The States may deal with it at their levels which might not be sufficient to tackle the possible consequences of using AI. In terms of technology (AI in particular), the world being a virtual village, steps need to be taken to evolve a comprehensive mechanism to deal with the legal aspect of AI.

15.8 SUGGESTIONS

- Clarification concerning the capacity of individual information, the preparation, and the degree of artificial knowledge to be utilized is to be made.
- It needs to establish an autonomous institution or body to examine the privacy policies while using AI of the business institution and certify its application.
- Constant vigilance of the business institutions to keep a check on the violation of these policies, and in case they violate, stringent penalties must be imposed upon them.
- The policies must be designed and applied so that those who do not wish to reveal their personal information must be exempted under them.
- The individual must have the right to negate or the right be forgotten to ensure the fundamental right to privacy in the arena of machines and AI.

KEYWORDS

- **artificial intelligence**
- **data protection**
- **deep learning**
- **human rights**
- **machine learning**
- **privacy laws**
- **right to privacy**

REFERENCES

1. Kerry, C. F., (2020). *Protecting Privacy in an AI-Driven World.* Brookings. https://www.brookings.edu/research/protecting-privacy-in-an-ai-driven-world (accessed on 07 September 2023).
2. Barrett, L., (2020). Ban facial recognition technologies for children—And for everyone else *Boston University Journal of Science and Technology Law* (Volume 26, No. 2). Available at SSRN: https://ssrn.com/abstract=3660118 (accessed on 07 September 2023).

3. 338 U.S. 25(1949).
4. 1963 AIR 1295.
5. 1975 AIR 1378.
6. AIR 1995 SC 264.
7. 1997 1 SCC 301.
8. 2005 1 SCC 496.
9. AIR 2010 SC 1974.
10. 2015 8 S.C.C. 735.
11. Buttarelli, G., (n.d.). Brussels, 30 November 2017 Keynote Speech. – Europa. from https://edps.europa.eu/sites/edp/files/publication/17-12-07_forum_europe_speech_en.pdf (accessed on 07 September 2023).
12. Manheim, K. M., & Kaplan, L., (2018). Artificial intelligence: Risks to privacy and democracy (October 25, 2018). *21 Yale Journal of Law and Technology, 106* (2019), Loyola Law School, Los Angeles Legal Studies Research Paper No. 2018–37, Available at SSRN: https://ssrn.com/abstract=3273016 (accessed on 07 September 2023).
13. *Mr Julian Knight MP Chair Digital, Culture and Media.* – ICO. (n.d.). https://ico.org.uk/media/action-weve-taken/2618383/20201002_ico-o-ed-l-rtl-0181_to-julian-knight-mp.pdf (accessed on 07 September 2023).
14. Rodrigues, R., (2020). Legal and human rights issues of AI: Gaps, challenges, and vulnerabilities. *Journal of Responsible Technology, 4*, 100005.
15. *Report of the Proceedings of the Online Expert.* – OHCHR, (n.d.). https://www.ohchr.org/Documents/Issues/DigitalAge/ExpertSeminarReport-Right-Privacy.pdf (accessed on 07 September 2023).
16. Mittelstadt, B. D., Allo, P., Taddeo, M., Wachter, S., & Floridi, L., (2016). The ethics of algorithms: Mapping the debate. *Big Data & Society, 3*(2), 205395171667967. https://doi.org/10.1177/2053951716679679 .
17. Osoba, O. A., & Welser, W. I. V., (2017). *The Risks of AI to Security and the Future of Work.* RAND Corporation. https://www.rand.org/pubs/perspectives/PE237.html (accessed on 07 September 2023).
18. *Unfairness by Algorithm: Distilling the Harms of Automated Decision-Making,* (n.d.). Future of Privacy Forum. https://fpf.org/blog/unfairness-by-algorithm-distilling-the-harms-of-automated-decision-making/ (accessed on 07 September 2023).
19. Bryson, J. J., Diamantis, M. E. & Grant, T. D., (2017). Of, for, and by the people: The legal lacuna of synthetic persons. *Artif. Intell. Law, 25*, 273–291. https://doi.org/10.1007/s10506-017-9214-9.

CHAPTER 16

Systematic Evaluation of Culture Biases in the Service Industry

CARI LEYSHON

CEO and Founder, CQify, Spain

ABSTRACT

The individual schema impacted key obstacles of diverse group interactions within a service-based organization. Findings included that shared individual biases can influence the effectiveness and development of artificial intelligence (AI). These strategies can be applied in using AI in the service industry. Key findings focused on understanding how an individual's unique schema includes multiple cultural identities to accept AI. The findings further confirm that the possibility of shifting perceptions can occur once individuals acknowledge their unconscious biases, stereotypes, and judgments. Diverse perceptions evaluated systemic biases within AI by applying a cultural intelligence (CQ) framework to engage the global customer. In addition, CQ training, with the precursor of understanding how the mind works, provides an opportunity for the developers of AI to mitigate their own biases. Using this business case indicates the importance of understanding individuals' unique reality and cultural lens prior to concluding the organizational challenge to identify efficient technical options to enhance the service sector. Individual biases collectively contribute to systemic biases. Therefore, it is necessary to initially take initiatives on the personal experience to inform the development strategy for AI.

Artificial Intelligence for Smart Technology in the Hospitality and Tourism Industry.
Vinod Kumar Shukla, Amit Verma, and Jean Paolo G. Lacap (Eds.)

16.1 INTRODUCTION

Social robots with a high level of perceived anthropomorphism, which determines the service industry's humanlike features, is gaining user acceptance [1]. Focusing on membership services within the enforcement industry, this study indicates a significant gap in artificial intelligence (AI) research that does not consider the benefits of applying a cultural intelligence (CQ) framework to reduce bias, stereotypes, and judgments of all stakeholders in the development of AI technology. The membership base includes one company from 18 different European countries. The role of culture needs to be considered to understand how it influences individual members' perceptions, which can continually contribute to misunderstandings and cause a reduction in stakeholder engagement or internal conflict.

Understanding how culture influences dialogic exchange was central to comprehending how individual perceptions are created by thought. These thoughts are fueled by their plethora of multicultural experiences creating a personal 'schema,' including judgments, biases, and stereotypes. The individual's culture also presented complex challenges when creating a trusting environment, considering how standard terms are interpreted differently with each cultural lens. Furthermore, a key indicator of user acceptance of AI is based on their perceived level of trust for the technology [1]. This inquiry explores the complexity of creating a trust based on the individual stakeholder's schema and cultural lens that needs to be identified to mitigate potential biases being perpetuated, either from the developers or the users. Initially, in the case of the enforcement network, the problem identified was believed to be how to increase member engagement. After further investigation, it was discovered that there was a misunderstanding between stakeholders on the individual perceived expectations of the service charter. Obstacles such as trust and misunderstandings hindered member collaboration with ineffective knowledge transfer, preventing the implementation of organizational objectives. To disrupt this non-productive cycle, critical reflection was necessary to profoundly investigate the process of how individuals can discover their unconscious biases by improving their CQ skills [2]. Around 85% of participants indicated that most people are not aware of their unconscious biases, existing stereotypes or automatic judgments that individuals have [3]. This significant finding could pose a potential risk to the development of AI if not all stakeholders consider how to implement a CQ model to mitigate bias. Suppose individual biases are not acknowledged in the development process. In that case, AI could be limited to a singular perspective which could potentially perpetuate discriminatory data based on labels, ethnic origin, pronouns, or any limiting view that is determined by the lens of the user, developer, or the system.

16.2 LITERATURE REVIEW

16.2.1 OBSTACLES OF DIVERSE COMMUNICATION

Obstacles of communication between the diverse memberships to be explored are investor analysis, management styles, individual schemes, and the role of culture, language, confidence, diverse associations, knowledge transfer, etc., as shown in Figure 16.1.

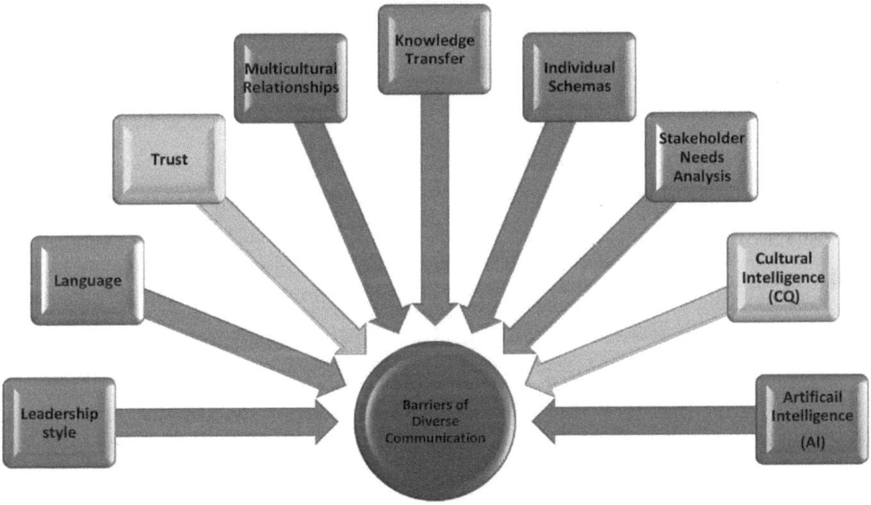

FIGURE 16.1 Identified topics of diverse communication.

The network members discussed the discrepancies in the interpretation of their service delivery standards due to their unique cultural lenses, which requires a new approach to synergize actions [4]. More recent research [5, 6] has contributed to the notion of "cultural intelligence (CQ)," that considers the influence of culture within the individual schema to inform communication strategies directed to customers with the advancements in AI technology [7].

Stakeholder analysis applying a CQ approach: A stakeholder analysis commonly focused solely on organizational objectives, although the advancement of technology now influences the way practitioners need to shift their focus on personalized culture [8–10]. Building trust between diverse stakeholders is crucial to becoming aware of how unconscious biases contribute to unchallenged individual perceptions [11]. This inquiry indicates how a stakeholders' perceptions could be contributing to misunderstandings between members [12]. The conscious acknowledgment of members' schemas is a crucial initiative to understand the root cause of the misunderstandings affecting relationships [13]. To engage the everyone in this discussion, it is necessary to facilitate an environment where there are high levels of trust between members to share their journey without fear of judgment from other members [11]. CQ skills acknowledge that we all share different realities and allows understanding and minimum possibilities of offending others' perspectives of others [14].

16.2.2 COMMUNICATION AND CULTURE

Every diverse stakeholder has a separate reality based on their cultural lens, which contributes to the complications of effectively exchanging information as intended. Cultural ques are becoming more challenging to navigate due to individuals having multiple cultural identities based on living and working in different countries from their place of birth [3]. Communication between individuals in diverse environments requires a conscious effort to avoid misunderstandings by applying many skills, including emotional and CQ [13, 15]. Misunderstandings are rooted within an individual's perception and unique cultural lens, which requires further investigation into schemas, developing trust, language, and sources of conflict as outlined in succeeding sections.

16.2.3 INDIVIDUAL SCHEMAS

A schema encapsulates an individual's biases, judgments, and beliefs that have been accumulated throughout their lives to create a unique interpretation of how the world works [15, 16]. Individual perceptions are influenced by their culture that determines how information is relayed, affecting distancing, tone of voice, and a multitude of essential cues that could be misunderstood [15, 17].

It is common to apply an affinity bias to seek people perceived to be most like ourselves. Conversely, it is important to get curious in diverse situations and not be threatened by differences [18]. Personal realities are based on our interpretations of our experiences in the world that are rarely challenged since they are accumulated selectively and influenced by education, culture, and various assumptions that contribute to misunderstandings.

Awareness of unconscious biases initiates the possibility for individuals to consider alternative information to shift their perceptions [19]. In addition, in his research, Adler confirms that misunderstandings are potentially rooted in the lack of openness to explore differences in perceptions due to unconscious bias [20]. To overcome these misunderstandings, individuals must be consciously aware of how the mind works, combined with the willingness to listen and authentically observe individuals in diverse settings deeply.

Furthermore, research confirms that an individual's ability to acknowledge their schema openly contributes to an authentic evaluation and consideration of multiple perspectives within an action research framework that would otherwise be neglected [21]. Change in perspectives can only occur after the individual consciously acknowledges their biases by considering information from diverse perspectives by understanding that every individual has a separate view of reality. The importance of soft skill development includes how individual attitudes can hinder or help teams collaborate and communicate to achieve objectives [22].

16.2.4 CREATING TRUST

Trust is defined as gauging whether commitments made between stakeholders are reliable. In addition, Kim and Van Dyne share that consistently across different cultures, trust and loyalty are both essential [23]. On the other hand, researchers argue that leaders can inadvertently dissipate trust by misusing language or actions that have different meanings based on the individuals' perceptions and cultural lens [24]. Agree that to facilitate a trusting environment, it is imperative to share their views, values, and perceptions to instigate a safe place for all stakeholders to participate [24]. Others emphasize that changing individual perspectives is not possible unless the individual is willing to reflect and challenge their beliefs that have been created selectively over their life experiences and are unique to how others view their reality. Trust will disappear once a misunderstanding is experienced [25]. Although, with an authentic and clear intention, it is possible to repair the relationship. Some researchers concur that authentic intention to build trustworthy diverse relationships supersedes having political correctness [26].

16.2.5 THE IMPORTANCE OF LANGUAGE

The advancement of technology is connecting the world and causing more opportunities for misunderstandings due to multiple languages and cultures, instigating reassessment of the need for a standard international language, such as English [3]. On the other hand, this is a delicate process to ensure that individuals fluent in different languages do not misinterpret this initiative as disregarding their cultural values, which could cause further conflict [7]. Isaacs confirms that misunderstandings often occur due to confirmation biases in action when the individual only selectively perceives information that confirms their existing belief systems and is not open to exploring different perspectives [27]. Other authors state that organizations fail due to cultural misunderstandings that are significantly higher in teams that do not have CQ training to dissect language [28]. Using native languages in smaller groups can support meaning-making between stakeholders using English [29].

Researchers Ochieng and Price, along with Kim and Van Dyne, emphasize that due to cultural differences, a leader must exhibit high levels of CQ to be more empathetic and build trusting environments to collaborate global teams [13, 23]. Soft skill development is a priority for leaders in this global economy. *Vulnerability* is an advanced leadership trait indicating a willingness to reflect on their schema to evaluate opportunities to adopt new perspectives. When leaders can shift their focus to understand every individual schema, it will facilitate an engaging stakeholder commitment towards objectives [30]. Alternatively, if leaders fail to invest time into individual teams, there could be a risk of disengagement [31].

16.2.6 CONFRONTING CONFLICT

The development of AI is propelling CQ skills as an essential management requirement as internal conflict and misunderstandings are causing multiple negative perceptions among key stakeholder groups [32, 33]. Misunderstandings require deep listening skills by an individual who possesses the openness and empathy to hear the other person [34] truly. Emotional intelligence utilized in AI has a positive response from users, although the further application of CQ may potentially contribute to user needs, especially in the service sector [35].

Deardorff agrees that misunderstandings between diverse stakeholders require a conscious ability to adapt and acknowledge the varying perceptions to overcome the source of conflict due to different interpretations of words or actions by applying CQ skills [24]. When the individual is experiencing intense emotions triggered by perceived conflict based on learned behavior, it requires an agile leader to diffuse the situation and guide a shift in perspective [33]. Improving CQ assists individual stakeholders to observe cultural cues and be open to different perspectives using ongoing reflection to seek a consensus to achieve key objectives [16, 36].

16.2.7 DEVELOPING INTERCULTURAL RELATIONSHIPS

The advancement of technology has allowed individuals to experience and gather information about different cultures and countries from anywhere in the [29, 37]. Bridging cultural differences are identified as a key management issue for businesses seeking to thrive and scale globally, which requires a new approach to creating synergy amongst diverse stakeholders [11]. Multiple cultural identities increase the complexities of programming AI without limiting labels of traditional descriptions of cultures that may no longer apply [38].

International migration is creating a third-generation culture where individuals live and work in different countries where either they or their parents were born [11]. Cultural differences are now considered a significant barrier without the agility of a leader with high CQ skills who can create a trusting environment to engage diverse perspectives [39]. Working within a diverse team requires participants to be consciously in the moment to deeply listen to others without judgment to truly understand their perspective while simultaneously acknowledging and challenging bias, judgments, and stereotypes with new information [24].

16.2.8 THE NATIONAL CULTURE AND ITS ROLE

National culture is a shared set of heterogeneous behaviors and actions deemed acceptable by groups of people to gauge individual acceptance

[40]. As per Seminal research adults' behaviors without exposing unconscious bias and adapting to external stimuli to be accepted within a chosen culture [41]. Research to date focuses on identifying cultural traditions and beliefs. However, limited studies have considered the impact of multiple cultural identities that minimizes the effectiveness of simply applying a stereotypical perception of understanding a specific group [26, 32, 40]. The unique combination of experiencing different cultures creates an individual perspective that is challenging to label or categorize, which can be a barrier for AI [42]. The application of AI has a meaningful opportunity to enable a rapid assessment of diverse perspectives and create a holistic solution without being vulnerable to human biases [35].

16.2.9 MINDFULNESS

Mindfulness is the ability to consciously be present to participate in a situation with awareness activities [43]. Further confirms how cultural ques differ from one's national culture, which requires personality traits that include a willingness and openness to consider alternative information that may contradict your schema. The neuroscience of how the mind works indicate that individuals feel their thinking with an unconscious belief, that their perception is the true, which neglects existing biases that can hinder relationships [44].

In 2005, Alon and Higgins shared that diverse teams offer a competitive advantage, although to prosper, the leadership team must acquire emotional and CQ skills [45]. A trusting learning environment is essential for diverse teams to clarify intended meanings and group rules [11]. If individual biases are left unacknowledged, their perceptions can influence misunderstandings between diverse stakeholders [22]. It is essential that individuals consciously work to acknowledge their own how their automatic judgments are discovered through ongoing reflection [16]. If participants are not willing to explore alternative information, it increases the unlikeliness of shifts in perception can occur [44].

16.2.10 LEADERSHIP

Leading diverse teams requires the agility to connect with individual stakeholders by navigating the varying perceptions to achieve global objectives [46]. In addition, Deardorff recommends the utilization of a lobbying to develop consensus on rules for participating in each unique team [2]. CQ skills provide the cultural agility required to engage diverse individuals with multiple cultural identities that could benefit the development of AI in the service sector [38]. Global leaders who commit to ongoing training to develop CQ skills can more effectively achieve team collaboration [47]. Use the word "edgewalker" to describe global leaders embracing this "new century leadership paradigm" to manage the complexities of leading diverse teams [3]. The existing research outlines the need to continually explore and gain knowledge about diverse leadership for companies to scale globally [36].

16.2.11 CULTURAL INTELLIGENCE (CQ)

Cultural intelligence (CQ) is an amalgamation of interpersonal skills and behaviors that provide the knowledge of how to adapt to diverse environments by navigating different cultures [6]. Globalization has propelled the need for all organizational stakeholders to develop CQ skills to overcome misunderstandings that disrupt collaboration [7]. The CQ framework is based on four areas identified as "cognitive, metacognitive, motivational, and behavioral" [5]. Crowne explains that individuals with CQ skills show empathy and consciously participate in dialog with others while simultaneously integrating perspectives [48]. Ramthun and Matkin argue that CQ skills are acquired with specific training and not simply traveling to different countries. When leaders make CQ a priority, it is possible to communicate with the individual while considering how multiple cultural identities restrict the application of generalizations [6, 49].

16.2.12 ARTIFICIAL INTELLIGENCE (AI)

Artificial intelligence (AI) is a collection of technology that incorporates four different types: mechanical, analytical, intuitive, and empathetic intelligence [1]. The empirical research to date is relatively new, and it is unknown if or how AI will be perceived as beneficial within the service industry [35]. Conversely, if AI can be applied and developed with CQ to transfer knowledge to increase the perceived anthropomorphism of user interactions, it could be a game changer [35, 38].

16.3 METHODOLOGY AND FINDINGS

Action research (AR) is an emergent process that requires the participation of stakeholders to contribute to the accumulation of all unique perceptions to create actionable knowledge [21]. In order to investigate the identified problem of this network, AR was selected for this inquiry using ongoing reflection and four action cycles [3, 16]. All 10 participants participated in each inquiry cycle to generate new information, which was applied to the next cycle to gather diverse perspectives using thematic template analysis. Each action research cycle focused on collecting data, analysis, and the researcher´s interpretations [21]. Figure 16.2 shows the research activities such as semi-structured interviews, focus groups and meeting facilitation, including the observation of participants.

Each cycle is built on the previous to develop the understanding within the network and address the research questions. The first cycle identified the problem based on stakeholders´ perceptions by contextualizing the literature for the network. The second cycle of inquiry gathered data using an online pre-test first following 30-minute semi-structured interviews with each participant [3]. Figure 16.3 shares an overview of topics identified relating to diverse communication and how each sub-topic (in orange) is explored [3].

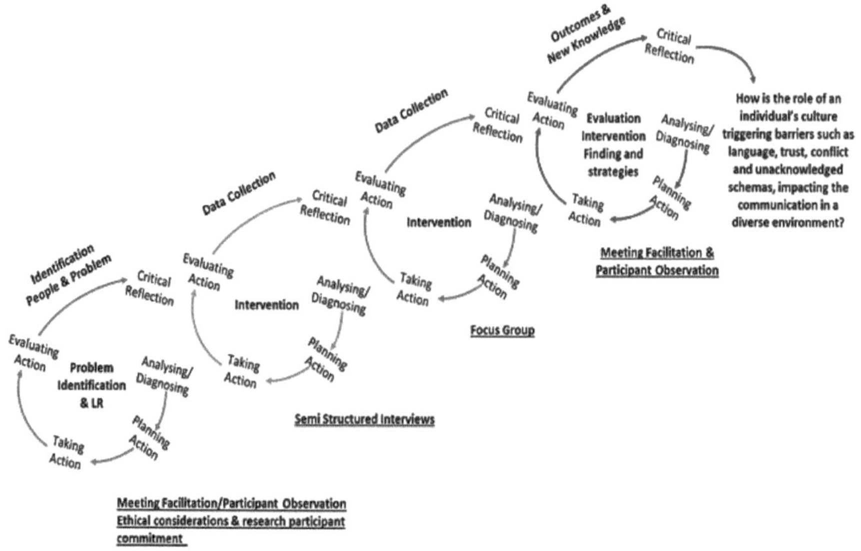

FIGURE 16.2 The four action cycles of inquiry [21].

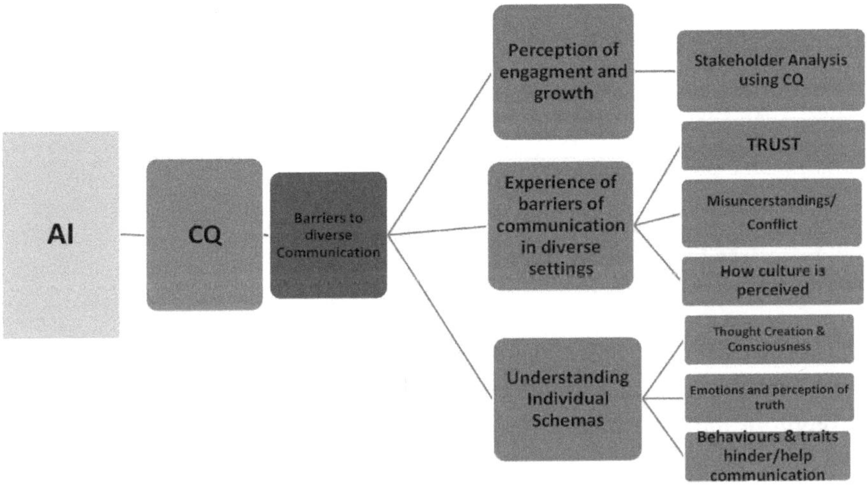

FIGURE 16.3 Difficulties to varied communication.

16.3.1 *CYCLE ONE – THE INQUIRY*

Dividing the members into small groups to deep dive into the identified problem of lack of member engagement could be due to miscommunication, lack of trust or the individuals' culture, which contributed to conflict and misunderstandings. To determine a consensus on the definition of inadequate communication within the group, three sub-topics were identified [3]:

i. What is a schema?
ii. Identifying experience of diverse dialog; and
iii. Collecting individual perspectives of what true engagement is.

The stakeholder discussion highlighted that cultural misinterpretations were occurring based on the individual's competence in using the English language. For example, participant A6 (2018) shared, "communicating in English is frustrating due to the difficulty and concentration to relay what I want to say which takes more time than in my language" [3].

16.3.2 *CYCLE TWO*

Online pre-test and semi-structured interviews: The 10 participants completed an online pre-test prior to the interview to shift their focus towards the network. A pattern emerged during the online semi-structured interviews as each individual shared examples of how their cultural lens contributed to their unique perception of situations. Participant A3 in 2018 stated, "As English is my second language, I look for nonverbal cues such as eye contact and body language to ensure mutual understanding" [3].

16.3.3 CYCLE THREE – FOCUS GROUP

The focus group provided a safe environment to authentically discuss and reflect on each participant's schema by consciously acknowledging their own biases and stereotypes that could hinder relationship building throughout the 90 minutes. As an example, participant A3 (2018) shared how they perceive communicating in a diverse environment. "I always perceive that the other person intends no harm, and I am open to them as a person, making sure my stereotypes of that culture do not hinder a potential relationship" [3]. During reflection, most participants agreed that our perceptions of experiences could be accumulated from our circle of influence and the culture we were raised in as a child [3]. The group consensus was that everyone innocently generally has stereotypes, judgments, and biases about different cultures. When our CQ skills are high, we can be open to nurturing relationships and shifting perceptions.

"Growing up, I was influenced to hate the same culture my parents did, although now as an adult my friendships include many people from the culture, confirming that our stereotypes can be changed if we are willing to reflect and acknowledge them" [3]. This new data shares that it is possible to shift individual perceptions, biases, and stereotypes when there is a willingness to acknowledge them consciously. If an individual continues to be reactive, emotional, and easily offended due to unconscious bias, misunderstandings can prevail. Alternatively, when participants are culturally intelligent and consciously aware of their schema, it sparks curiosity about different perspectives [3]. The data generated by this inquiry was first linked with the literature, then thematic template analysis and interpretation were used by the researcher [3, 50]. A fundamental aspect of this research journey was the importance of reflection in action, which enabled exposing my own unconscious biases and opened up new learning opportunities [44].

16.3.4 CYCLE 4 MEETING PARTICIPATION

The stakeholder analysis of the 10 participants in this research inquiry indicated how perception is influenced by culture. When the unconscious bias is uncovered, and new information can be considered, obstacles subside, contributing to conflict and lack of trust. The process of creating a trusting environment is based on a foundation of high levels of trust created by the ongoing development of CQ skills within all stakeholders involved [3].

Once the participants became consciously aware of their stereotypes and biases, it was possible to shift perspectives, even when information contradicted their initial perception created over the individuals' lifetime. The role of an individual's culture profoundly influences how dialog is received and judged based on the findings. This inquiry instigated a desire to gain further training to improve individual levels of CQ to perform better within diverse environments.

Language and the individual's competence level in English is a real obstacle between stakeholder communications. CQ skills do not equate to how much participants have traveled or experienced different cultures. Another interesting finding was the importance of the social part of the meeting itinerary where participants could communicate in a relaxed environment, build trust, and respect that spilt over to the official part of the meetings. The discovery of creating a customer service charter that specifically collected key terms and actions completed by the membership, to overcome potential misunderstanding of common terms that hold different meanings in different cultures [3]. In addition, the membership was able to collaborate more effectively as CQ skills improved combined with conscious awareness to overcome stereotypes and judgments that were historically interfering with effective communication, as outlined in Figure 16.4.

16.4 DISCUSSION

- **Research Question 1:** How do participants perceive the benefits of membership?

  Exploring this initial research question, the researcher could build trust and communication with the participants. The purpose

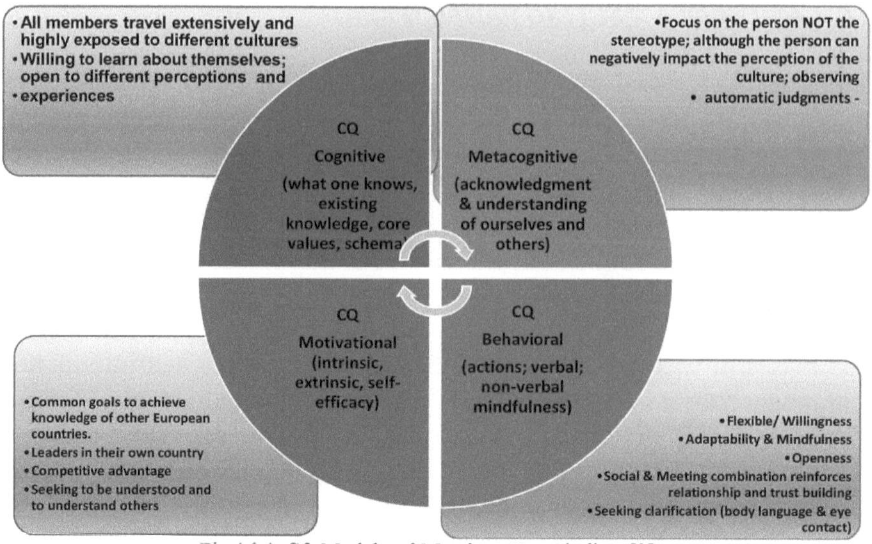

FIGURE 16.4 CQ model and members key findings [3].

of the network was to offer clients access to enforcement in the 18 different countries of their membership, achieving a competitive advantage within the localities. During this initial cycle, it was highlighted that misunderstandings could displace the overall benefits. As the research cycles progressed, it was discovered that the culturally determined email response times varied. For example, it is common to respond to an email within a week versus another who expect a same-day response in one country. This individual subconscious judgment festered into some members experiencing a reduction of trust in the system to respond to their clients in their perception of a prompt response. This finding is significant since it exposed how diverse cultures induces an array of interpretations that need to be clarified to gain trust and collaboration [3].

This unexpected finding exposes the source of potential misunderstandings resolved by implementing a member service charter with specific terms and actions defined collectively by the members. Understanding each member's perception by improving CQ skills helps to acknowledge unconscious biases that can clear the pathway towards achieving collective action [3].

- **Research Question 2:** What obstacles are perceived during diverse communication between members?

The consistent feedback from participants was the acknowledgement and personal accountability of understanding the importance of deep listening while being open to discovering new information from people from different cultures to gain multiple perspectives [3, 51]. The discovery that there were 18 different definitions or interpretations of the term 'trust′ initiated the further investigation of other terms and actions used within the network context [3]. Learning CQ skills is an ongoing process that is best nurtured using a coaching leadership style to engage stakeholders' long term [5, 6, 49]. It was interesting to observe how the exposure to different cultures caused the members to adapt mindfully in each situation to build a deep level of trust [52]. On the other hand, these soft skills improved with the individual acknowledgment of their schema within a supportive environment [3].

A key point within this research was understanding that language is identified as an obstacle when trying to relate specific thoughts or feelings using English as a second language, contributing further to misunderstandings without CQ skills. Specifically, the participant (C1, 2018) shared the advantage of using multiple languages when required to ensure the meaning is translated as intended could be helpful [3]. This should be applied mindfully not to alienate other members. If there is uncertainty using verbal language in diverse situations, non-verbal cues are essential, as indicated within the literature [15, 17]. Understanding the obstacle of language assists the group to become consciously aware of the possible misinterpretation of meanings.

- **Research Question 3:** Does the individual schema positively or negatively influence multicultural relationships versus same culture relationships?

The data collection process uncovered the potential difference in perception of how members experience building relationships with people from the same alternative cultures as themselves. After a deeper investigation, it was revealed that communication obstacles within diverse settings could be reduced between members who share trust and individual acknowledgement of biases, judgments, and stereotypes [53]. Arguably, suppose there is no willingness to improve their level of CQ and fail to pursue the acknowledgement of existing biases, judgments, and stereotypes. In that case, no change can occur with their perception of reality remaining the same [54].

The evidence from the data suggests that if the participant lacks CQ skills, this could block the ability to form relationships with people from different cultures [45]. Culture is at the heart of informing individual perceptions should be significant priority for organizations to maximize the value of diverse perspectives [22].

- **Research Question 4:** Do cultural stereotypes effect the participant's perception of communication?

The disruption of effective communication between cultures could be rooted within each individual's schema. Stereotypes of people and cultures are common, although these are mainly held within our subconscious programming from the day we were born and through life experience. The researcher engaged the participants on a journey to bring these stereotypes into a trusting environment to explore examples collectively. Surprisingly, results showed they no longer hold value in most cases [3]. Once stereotypes were discussed, the data confirmed how this influence the individual perceptions of different cultures and can only be altered with recognition. As each research cycle progressed, it was imperative for the researcher to critically reflect on how or if their own biases were impairing the interpretations of the participants' perceptions. The researcher guided each participant to explore the possibility of shifting their perceptions once the schema was acknowledged in the disruptive process outlined in Figure 16.5.

The data generated confirmed how the participants did shift their perceptions by acknowledging the role of culture within their schema while understanding how to develop their CQ skills further and using ongoing critical reflection [53]. The findings also suggest that culture innocently feeds the individual perception of reality without any judgment of being correct or not, as shared by Oliveira [15]. All participants' feedback and perspectives were validated by existing literature to navigate the obstacles of diverse communication using the research questions as a map [3].

Initially, the identified problem was member engagement, although the findings further clarify how misunderstandings due

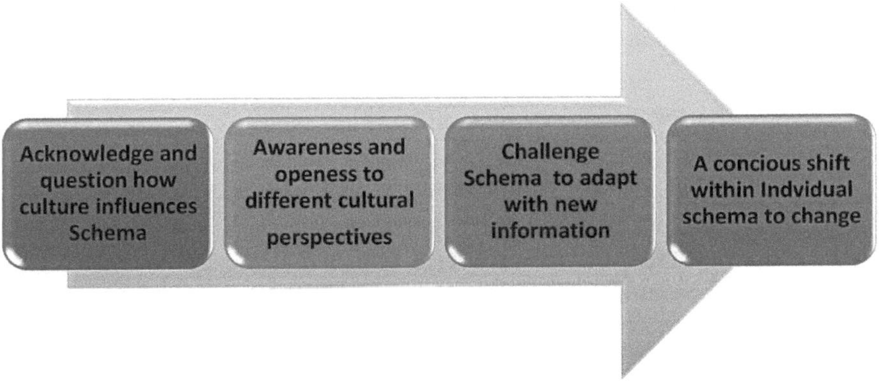

FIGURE 16.5 Individual schematic adjustment process [3].

to cultural differences were the source of hidden conflict based on alternative perspectives and expectations. Surprisingly, it was discovered that individuals perceived themselves as being open to considering different perspectives, indicating that the barriers were at an unconscious level [36, 55].

Vulnerability can be perceived as a weakness, especially in a professional setting [56]. The participants shared their stereotypes, making them vulnerable to judgment within the group, emphasizing the facilitator's skill to create a trusting environment and keep the sharing within the context of the discussion. Guiding the participants through the discovery of acknowledging their schema was perceived as an emotional risk that was overcome by developing CQ skills and necessary personality traits required to challenge existing beliefs for the benefit of this study. The role of culture and the perception of different cultures has a powerful influence on individual decision making as verified by "participant C2, who shared how he perceived his culture is not respected highly enough for him to be voted on the board" [3].

Once this stereotype is acknowledged, it is possible to shift this perception by challenging what information this is based on and introducing alternative perspectives [36]. Culture is ingrained into our beliefs and unconsciously informs our perception of acceptable behavior, causing blind actions and creating opportunities for misunderstandings blocking organizational objectives. It was observed how participants shared that they consider themselves open to engagement with all types of cultures and then unexpectedly used a stereotype randomly within the data collection process [3].

As the facilitator, it is imperative to possess a high level of CQ to observe cultural cues to create a trusting environment and relationships with the members to share perceptions authentically without fear of judgment or consequence [38]. In addition, active dialogic deep listening with a conscious interest to receive the information is helpful to achieve an effective dialog exchange [11, 39]. A key experience during the study was the consolidation of perspectives once individuals understood how their schema influenced their perceptions of others and recognized that it is possible to change our perceptions. The interpretation of the findings by the researcher indicated that during the study, the participant's level of CQ improved consistently throughout each cycle of inquiry [3].

Although it is essential to highlight that by using critical-dialogic empathy without high levels of trust, participants would not share deeply guarded insights into their schema, potentially fueling denial or other common natural defense mechanisms [57]. Critical dialogic empathy used in the focus group provided a safe environment to disclose personal stereotypes without offending other participants in the form of collective mindfulness [3, 43].

Introducing potentially emotive discussions into a group required a leader with high levels of CQ to mitigate [3, 58] mindfully. Navigating through vulnerable discussions required ongoing reflection, balancing validation with empathy to avoid misunderstandings during the data collection [58, 59]. The researcher´s interpretation that as participants began to acknowledge their stereotypes, judgments, and beliefs, it became apparent how influencing culture is [24, 40].

This process of self-discovery combined with ongoing reflection to consciously be aware of our decision making has enabled a change in perceptions [16, 44]. A relevant finding discloses that people are experiencing different cultures than their country of birth due to migration, which means individuals have multiple cultural identities [3, 36]. It is important to note that being exposed to different cultures does not equate to your CQ skill level [52].

This finding reinforces that it is most effective to understand the individual′s culture [3] initially.

- **Research Question 5:** What are the participant's perceptions of commitment and development prospects of the organization?

The focus for the first meeting is to explore the different perceptions of the members and about the problem they are experiencing in the context of the network. The purpose of the initial meeting was to identify the perceived problem the network was experiencing. Initially, the consensus among members was that engagement was a priority, which evolved as communication and then English competency levels. During the meeting, members acknowledged awareness of different cultures although not recognizing the power of unconscious bias and how perceptions are informed. By placing the members in small working groups, it was possible to identify how the perceived problem is more profound than acquiring communication skills. Reverting to the literature, the researcher gained insight into various obstacles that hinder diverse stakeholder collaboration, which guided the approach and methodology of the study [3].

Reflecting on the findings, it was apparent how the participant's observations have altered between the first and fourth research cycle. These findings validate the impact of improving individual levels of CQ skills to influence their schema. Through the conscious and mindful process of challenging individual perceptions, it was discovered that it is likely to modify previous faiths, prejudices, and stereotypes to overcome obstacles preventing implementing organizational strategies [3]. All members attended the final meeting, showcasing the difference in behaviors between the study participants and the other members. CQ training is only beginning, and positive effects are apparent, although it must continue for sustainable change to be embedded.

The researcher observed the differences in CQ skill levels of all members, considering how the data was shared. Some were not previously aware of any differences between cultures as opposed to others who prefer to avoid diverse situations if possible [3]. Another interesting finding is that the more personal information and social interactions stakeholders shared, the more perceived trust [3]. Cultural curiosity generates deeper insight into different cultures, reducing fear of the unknown or misinterpretation [17, 57]. The members displayed ethno-relative behavior by showing more agility in the diverse setting [3, 60].

During the last meeting, all stakeholders' perceptions on the definition of terms and expected service standards were amalgamated to create the first moral circle within the context of the network [24]. The application of the moral circle concept diluted the possibility of ongoing misunderstandings as this information provided a clear outline that the members can refer to stimulate engagement and trust. The hidden reason for misalignment issues was the confusion and expectations on the appropriate response times to online requests. The moral circle has had a profound positive influence on member engagement since all members now clearly understand and agree on the service terms created based on the collective input. The key takeaway is to ensure that all stakeholders participate in the moral circle so all perceptions can be integrated and applied to the moral circle [3].

In further findings, it was confirmed that regular membership communications are necessary, and when a member's failure to contribute, it could be potentially disruptive to the network benefits. Inclusive leadership by the facilitator who has achieved a high level of CQ skills understood the impact of the individual's culture [3]. Upon reflection, the facilitator can't adapt themselves to appeal to the group due to individuals having multiple cultural identities. Therefore, members with CQ skills are prone to seek understanding as opposed to focusing on what could be perceived as offensive [3, 24, 31]. To create a trusting environment, the facilitator discovered that acknowledging the innocence behind biases and that everyone has them opened the dialog for members to share their own [61].

Based on the findings, it became apparent that using an insider researcher inquiry posed a strong advantage with the existing relationship with the stakeholders fast-tracked the trust-building process when the investigator is taken as a part of the team. Upon reflection, it could have been advantageous to follow up with an additional interview with the participants for more feedback on their perceptions of new insights after the last meeting [3]. On the other hand, participants shared their personal experiences of how they perceived their schema shifted during the research process, which is a positive confirmation to continue the collective initiative of CQ skills training [3].

16.5 SUMMARY

16.5.1 OBSTACLES OF CULTURE WITHIN INDIVIDUAL´S SCHEMA

The cultural lens used to create the individual's perception of reality in diverse situations can potentially contribute to misunderstandings between members. Understanding how culture is central to conducting a stakeholder analysis with a leadership approach that utilizes CQ skills is key to building a trusting environment [3]. Managing diverse teams requires cultural competence on the part of all member's intelligence skills as it is impossible for one person to adapt to a different culture. When every member is accountable to participate in diverse situations consciously, it is possible to disrupt obstacles that may interfere with effective knowledge transfer [3]. Once trust is established between members, it is possible to acknowledge unconscious bias, stereotypes, and automatic judgments to disrupt the pattern and allow new information to shift perspectives. Diverse perspectives improve decision-making and reduce individual bias. With realization, it is possible to change one's perception of reality or circumstances. CQ goes deeper than just an awareness of specific customs and traditions by triggering a curiosity to learn about differences. Improving members' CQ skills while simultaneously acknowledging their schemas produced a collaboration towards overcoming the obstacles culture could cause [3].

The researcher applied a social constructionist lens to collect the diverse perspectives through the cycles of inquiry. Based on a positivist perspective, the analysis would be limited to focusing solely on the empirical data, neglecting the value of applying subjective input. Participants shared the value of experiencing personal growth through each action cycle of the research process instigated by understanding the role of culture within the schema [3].

The process of acknowledging unconscious bias requires courage and the ability to be vulnerable within a trusting environment led consciously by a leader with advanced CQ skills. This study concluded that it is possible for individuals who were previously unaware of their unconscious biases to change their perceptions regardless of their culture. The human element in this process is fundamental to navigating the participants' emotions, building trusting relationships, and observing critical non-verbal cues that AI would not. Trust between all stakeholders is essential to the successful outcomes of this type of intervention. The skillful integration of exposing stereotypes without judgment to navigate potentially emotive discussions by retaining focus within the network context requires deep listening and group consensus [3].

16.5.2 KNOWLEDGE TRANSFER

The theoretical labels framed the inquiry using an online pre-test, interviews, focus group to conduct 4 action research cycles. The complexity of the problem was exposed on initial inquiry as existing research was sifted through to set up the data collection process. Besides the clear objectives, the journey was messy, and mindful facilitation was required to establish a trusting environment.

Each participant is already familiar to me as an insider researcher and I could dig deeper into their schemas based on the existing trust, which was fundamental to the inquiry's credibility. It was necessary to transfer the expertise and competencies to enhance better CQ. In attempting to overcome misunderstandings with diverse stakeholders, the researcher required numerous strengths to motivate each individual to recognize their configuration. The data confirmed that with individual acknowledgement of their biases, stereotypes, and judgments were, the personal change could arise. The literature relates with the challenges that communication barriers cause [3].

Culture has the potential to be associated to the compass guiding our core values and beliefs [36]. Disregarding the cultural impact will persist to stymie trust, resulting in conflict and resulting in a lack of collaboration of diverse stakeholders [3, 6, 23, 62]. All the aforementioned themes were inextricably linked to form a set of skills that included trust in order to create an intercultural institution with high CQ relevant parties and leaders who are "edgewalkers" that will flourish in this international economy [3, 24, 51, 60].

16.5.3 KEY OUTCOMES: EVALUATING THE HUMAN INPUT AND AI

The findings of this study confirm that the multiple cultural identities of stakeholders minimize the capability of AI to use labels that generalize or categorize humans effectively. The outcomes of the four research cycles dissect the process of individual and organizational learning while emphasizing the role of trust [3]. Understanding the unique obstacle of how culture peppers individual perception could be helpful in the development of AI within the service sector:

16.5.4 ACTIONABLE KNOWLEDGE APPLIED BY PARTICIPANTS [3]

- Experiential small group sessions and one-to-one interaction led to a journey of individual acknowledgement of their schemas by disclosing existing stereotypes in a trusting environment which is fundamental for change of perceptions to occur.
- They are gaining new knowledge about how CQ skills are not dependent on exposure to multiple cultures and go deeper than just cultural differences.
- To improve diverse stakeholder collaboration, using small working groups enables all participants to contribute to key decisions.
- CQ training unleashes the unique opportunities to consciously engage with curiosity about differences instead of being fearful or avoiding them.
- Understanding how authentic critical reflection makes shifting perspectives and existing stereotypes possible.
- Competence levels in English vary between stakeholders, which requires a higher reliance on non-verbal cues and the option to clarify meanings in different languages without alienating group members.
- Organizational change requires a leader with high CQ skills to navigate potentially emotive situations to overcome conflict and misunderstandings.
- Trust is an invisible force that is established in the process of relationship-building with authentic dialog and when stakeholders can engage socially.

- Diverse stakeholders can improve communication by developing more patience with a mindful approach.
- Mobility and technological advancements create global citizens with multiple cultural identities challenging traditional cultural generalizations.

Introducing learning sets and the virtuous circle improved decision making, collaboration engagement and productivity of members. The perceived inconvenience of the additional time of interviewing individual stakeholders prior to meeting proves to fast-track clarity and consensus with the ability to identify potential misunderstandings [3] quickly. Reflecting on this research journey, I am consciously aware of my own development from increasing my CQ skills in my role as an insider action researcher [3, 48]. My interaction with all participants during the four cycles presented an opportunity to gain a good grasp of their schemas while evaluating and challenging my own [3].

Before this research project, I assumed we all saw the world from a similar perspective. I had never understood that existing stereotypes could be changed when we become open and curious to learn that there are multiple things, view, and live in the world. In particular, the most profound experience from my perspective was the courage and vulnerability of the participants to authentically put a voice to their unconscious biases, understanding we all have them. Conflict is rooted in misunderstandings and a lack of initiative to consider alternative perspectives from your own. Working and socializing in diverse environments requires a multitude of skills, mindfulness, and sensitivity, which I never considered at such a deep level in the past [3].

There is a high risk of sabotaging relationships by missing fundamental cultural cues due to allowing yourself to be blindly guided by unconscious bias and automatic judgments. People are unique, with multiple cultural identities that make historical labels obsolete. This specific organization within the service industry can inform AI developers of the importance of acknowledging bias and centralizing the responsibility of society.

16.5.5 AI AND CQ LEADERSHIP INITIATIVES

AI could positively contribute to the service industry if there is a focus to reduce individual societal and systemic bias. Research to date has focused on accumulating synthetic data to provide the data set with a diverse range of ages, genders, and ethnicities combined with several different variables. Wherein AI developers are not considering the individual biases of the programmers. Therefore, in the initial development of AI, it is recommended that the developers consider their level of CQ by acknowledging unconscious biases that contribute to automatic judgments. These stereotypes create their perception of reality. Once this understanding is achieved, the development team would consider multiple perspectives that would produce technical applications that are perceived as trustworthy and engaging to the diverse user. Equally, it is essential to cleanse and acknowledge systemic biases that could interfere with the development of the technology by perpetuating unconscious bias, as shown in Figure 16.6.

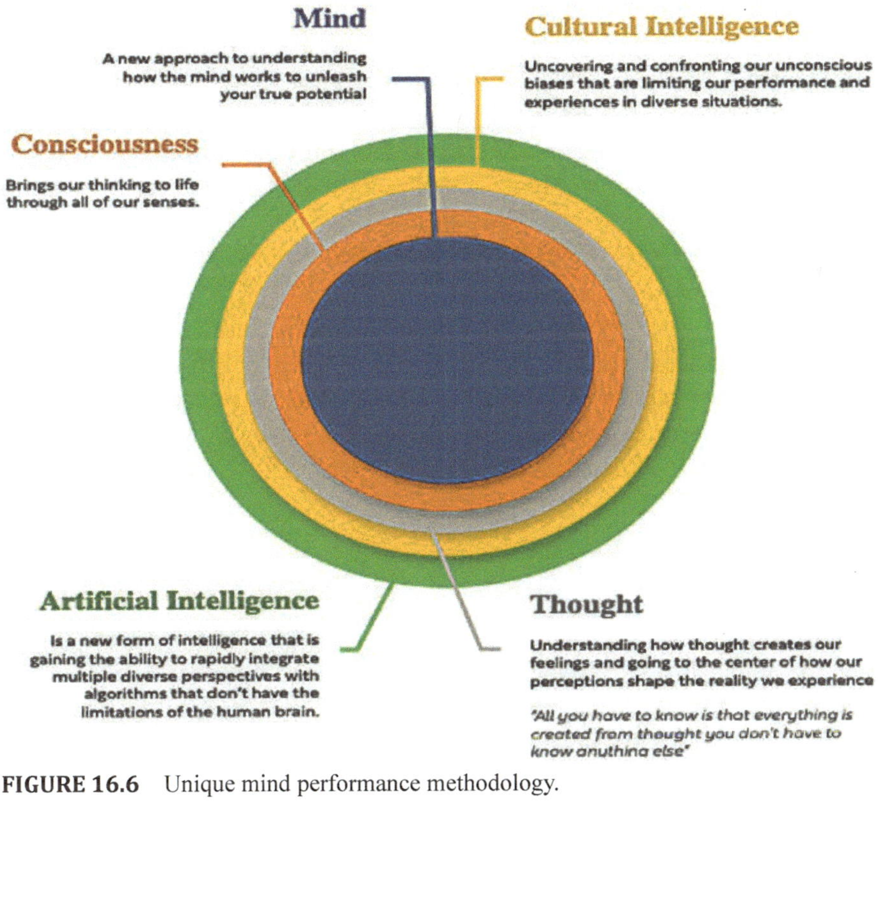

FIGURE 16.6 Unique mind performance methodology.

Understanding that each individual has separate realities based on their cultural lens is still a significant oversight in the development of AI. Focusing on social AI to emulate human feelings and behaviors, it is important to collect data from diverse stakeholders to maximize adaptability. If this step is disregarded, AI could then perpetuate stereotypes, judgments, and individual bias. Research on the development of social AI does discuss the importance of emotional intelligence to show the necessary empathy humans interact with. The literature gap indicates that AI developers have not applied the soft skills associated with CQ to advance user engagement of AI.

AI can sabotage efficiencies with the service sector if there is a lack of CQ skills. The positive is that CQ skills are constantly improving once the individual is aware or acknowledges unconscious bias and is open to diverse perspectives. The process of acknowledging all forms of both individual and systemic biases could provide a clear implementation strategy for the continued development of AI in the service industry. It is important not to make assumptions on the labels or process without considering individual biases from each person involved in the development of AI technology. Without the consideration of the multiple cultural identities of the user, there could be a risk of producing inefficient processes that would limit the adaptability or scalability of the AI.

Further research is required to implement CQ training that integrates the understanding of how the mind works in an intense program for individual AI developers to disrupt the transfer of their own bias, stereotypes, and judgments. In addition, this process could potentially inform how further to mitigate the users, societal, and systemic bias. With this understanding and training integrated with AI technology, the future of efficient applications in the service industry are potentially endless.

16.6 CONCLUSION

If additional time were allocated, it could have provided a deeper investigation regarding the impact CQ skills impacted both stakeholders and the network [3]. Furthermore, individual interviews with all stakeholders could have potentially disclosed additional perspectives compared to the study participants to understand their varied experiences.

KEYWORDS

- **artificial intelligence**
- **change management**
- **communication**
- **cultural intelligence**
- **future marketing**
- **global consumer**
- **systemic bias**

REFERENCES

1. Chi, O. H., Denton, G., & Gursoy, D., (2020). Artificially intelligent device use in service delivery: A systematic review, synthesis, and research agenda. *J. Hosp. Mark. Manag, 29*(7), 757–786. doi: 10.1080/19368623.2020.1721394.
2. Deardorff, D. K., (2009). *The Sage Handbook of Intercultural Competence*. Thousand Oaks, Calif: Sage Publications.
3. Cari, L. L., (2019). H00020329_MAR2019-DBA-Thesis-The Case of a European Network.pdf. https://livrepository.liverpool.ac.uk/3090922/1/H00020329_MAR2019-DBA-Thesis-The%20Case%20of%20a%20European%20Network.pdf (accessed on 07 September 2023).
4. Gratton, L., & Erickson, T. J., (2007). Eight ways to build collaborative teams. *Harvard Business Review*. [Online]. Available: https://hbr.org/2007/11/eight-ways-to-build-collaborative-teams (accessed on 07 September 2023).
5. Earley, P. C., & Ang, S., (2003). *Cultural Intelligence: Individual Interactions Across Cultures*. Stanford: Stanford University Press.
6. Ott, D. L., & Michailova, S. (2018). Cultural intelligence: A review and new research avenues. *International Journal of Management Reviews*, *20*(1), 99–119, https://doi.org/10.1111/ijmr.12118 (accessed on 07 September 2023).
7. David, C. T., & Kerr, I., (2017). *Cultural Intelligence: Surviving and Thriving in the Global Village* (3rd edn.). Oakland, CA: Berrett-Koehler Publishers, Inc., 2017.
8. Bryson, J. M., (2004). What to do when Stakeholders matter: Stakeholder identification and analysis techniques. *Public Manag. Rev.*, *6*(1), 21–53. doi: 10.1080/14719030410001675722.
9. Reed, M., et al., (2009). Who's in and why?: A typology of stakeholder analysis methods for natural resource management. *J. Environ. Manage.*, *90*(5), 1933–1949. doi: 10.1016/j.jenvman.2009.01.001.
10. Kivits, R. A., (2011). Three-component stakeholder analysis. *Int. J. Mult. Res. Approaches*, *5*(3), 318–333. doi: 10.5172/mra.2011.5.3.318.

11. Harvey, M. G., & Griffith, D. A. (2002). Developing effective intercultural relationships: The importance of communication strategies. *Thunderbird International Business Review*, *44*(4), 455–476.
12. Duncan, T., & Moriarty, S. E., (1998). A communication-based marketing model for managing relationships. *J. Mark.*, *62*(2), 1–13. doi: 10.2307/1252157.
13. Ochieng, P. E., & Price, A., (2009). *Managing Cross-Cultural Communication in Multicultural Construction Project Teams: The Case of Kenya and UK.* Social Science Research Network, Rochester, NY, SSRN Scholarly Paper ID 2621174. [Online]. Available: https://papers.ssrn.com/abstract=2621174 (accessed on 07 September 2023).
14. Lindholm, C., (2008). Handbook of cultural psychology Shinobu Kitayama Dov Cohen. *J. Anthropol. Res.*, *64*, 569, 570. doi: 10.1086/jar.64.4.20371296.
15. De Fatima, O. M., (2013). Multicultural environments and their challenges to crisis communication. *J. Bus. Commun.*, *50*(3), 253–277. https://doi.org/10.1177/0021943613487070.
16. Alvesson, M., & Sköldberg, K., (2017). *Reflexive Methodology: New Vistas for Qualitative Research.* SAGE.
17. Neuliep, J. W., (2017). *Intercultural Communication: A Contextual Approach* (7th edn.). Los Angeles: SAGE Publications, Inc.
18. Fine, M. G., (1991). New voices in the workplace: Research directions in multicultural communication. *J. Bus. Commun.*, *28*(3), 259–275. https://doi.org/10.1177/002194369102800305.
19. *Values, Schemas and Norms in the Culture-Behavior Nexus: A Situated Dynamics Framework* | Columbia Business School Research Archive. https://www8.gsb.columbia.edu/researcharchive/articles/24426 (accessed on 07 September 2023).
20. Adler, N. J. (2003). Communicating across cultural barriers. *Interkulturelle Kommunikation: Texte und Übungen zum interkulturellen Handeln. Sternenfels: Wissenschaft & Praxis*, 247–276.
21. Coghlan, D., (2010). Seeking common ground in the diversity and diffusion of action research and collaborative management research action modalities: Toward a general empirical method. *Res. Organ. Change Dev.*, *18*, 149–181. doi: 10.1108/S0897-3016(2010)0000018009.
22. Siakas, K. V., Georgiadou, E., & Balstrup, B., (2010). Cultural impacts on knowledge sharing: Empirical data from EU project collaboration. *VINE*, *40*(3, 4), 376–389. doi: 10.1108/03055721011071476.
23. Kim, Y. J., & Van, D. L., (2012). Cultural intelligence and international leadership potential: The importance of contact for members of the majority: International leadership potential. *Appl. Psychol.*, *61*(2), 272–294. doi: 10.1111/j.1464-0597.2011.00468.x.
24. Berardo, K., & Deardorff, D. K., (2012). *Building Cultural Competence: Innovative Activities and Models*. Stylus Publishing, LLC.
25. Jahansoozi, J., (2006). Organization-stakeholder relationships: Exploring trust and transparency. *J. Manag. Dev.*, *25*, 942–955. doi: 10.1108/02621710610708577.
26. Dumitru, V., & Cozman, D., (2012). The relationship between stress and personality factors. *Hum. Vet. Med.*, *4*. .

27. William N. Isaacs (1999). ISAACS, dialogic_leadership.pdf. http://www.dialogos.com/wp-content/uploads/2020/10/dialogic_leadership.pdf (accessed on 07 September 2023).
28. House, R. J., Dorfman, P. W., Javidan, M., Hanges, P. J., & De Luque, M. S., (2013). *Strategic Leadership Across Cultures: The Globe Study of CEO Leadership Behavior and Effectiveness in 24 Countries* (1st edn.). Thousand Oaks, California: SAGE Publications, Inc.
29. Poncini, G., (2003). Multicultural business meetings and the role of languages other than English. *J. Intercult. Stud.*, *24*, 17–32. doi: 10.1080/07256860305790.
30. Men, L. R., & Stacks, D., (2014). The effects of authentic leadership on strategic internal communication and employee-organization relationships. *J. Public Relat. Res.*, *26*(4), 301–324. doi: 10.1080/1062726X.2014.908720.
31. Shanna, L., Zachary, S., & Caroline, H., (2015). *Investigating the Effect of Anxiety, Uncertainty and Ethnocentrism on Willingness to Interact in an Intercultural Communication.* https://journals.sagepub.com/doi/abs/10.1177/0022022114555762 (accessed on 07 September 2023).
32. Richard, D. L., (2016). When-Cultures-Collide.pdf. http://www.utntyh.com/wp-content/uploads/2011/11/When-Cultures-Collide.pdf (accessed on 07 September 2023).
33. Hammer, M. R., (2005). The intercultural conflict style inventory: A conceptual framework and measure of intercultural conflict resolution approaches. *Int. J. Intercult. Relat.*, *29*(6), 675–695. doi: 10.1016/j.ijintrel.2005.08.010.
34. Livermore, D. A., (2015). *Leading with Cultural Intelligence: The Real Secret to Success* (2nd edn.). New York: American Management Association.
35. Collins, R. *How Artificial Intelligence Will Transform Human Thinking*. Management-Issues.com. https://www.management-issues.com/opinion/7362/how-artificial-intelligence-will-transform-human-thinking/ (accessed on 07 September 2023).
36. Knoll, M., Meyer, B., Kroemer, N., & Schröder-Abé, M., (2015). It takes two to be yourself: An integrated model of authenticity, its measurement, and its relationship to work-related variables. *J. Individ. Differ.*, *36*, 38. doi: 10.1027/1614-0001/a000153.
37. Mark, C. G., & Warren, J. K., (2016). *Global Marketing*. https://www.biblio.com/global-marketing-by-mark-c-green-warren-j-keegan/work/87216 (accessed on 07 September 2023).
38. Articles by David Livermore. https://www.management-issues.com/david-livermore/ (accessed on 07 September 2023).
39. Booher, D. E., & Innes, J. E., (2002). Network power in collaborative planning. *J. Plan. Educ. Res.*, *21*(3), 221–236. doi: 10.1177/0739456X0202100301.
40. Hofstede, G., (2006). What did globe really measure? Researchers' minds versus respondents' minds. *J. Int. Bus. Stud.*, *37*, 882–896. doi: 10.1057/palgrave.jibs.8400233.
41. Hall, E. T., & Whyte, W. F., (1960). Intercultural communication: A guide to men of action. *Int. Exec.*, *2*(4), 14, 15. doi: 10.1002/tie.5060020407.
42. Manyika, J., Silberg, J., & Presten, B., (2019). What do we do about the biases in AI? *Harvard Business Review*. [Online]. Available: https://hbr.org/2019/10/what-do-we-do-about-the-biases-in-ai (accessed on 07 September 2023).

43. Vogus, T. J., & Welbourne, T. M., (2003). Structuring for high reliability: HR practices and mindful processes in reliability-seeking organizations. *J. Organ. Behav.*, *24*(7), 877–903. doi: 10.1002/job.221.
44. Rigg, C., & Trehan, K., (2008). Critical reflection in the workplace: Is it just too difficult? *J. Eur. Ind. Train.*, *32*, 374–384. doi: 10.1108/03090590810877094.
45. Alon, I., & Higgins, J. M., (2005). Global leadership success through emotional and cultural intelligences. *Bus. Horiz.*, *48*(6), 501–512. doi: 10.1016/j.bushor.2005.04.003.
46. Zander, L., & Butler, C., (2010). Leadership modes: Success strategies for multicultural teams. *Scand. J. Manag.*, *26*, 258–267. doi: 10.1016/j.scaman.2010.06.002.
47. Ozguler, I., (2016). Increase the projects' success rate through developing multi-cultural project management process. *Procedia – Soc. Behav. Sci.*, *226*, 236–242. doi: 10.1016/j.sbspro.2016.06.184.
48. Adair, W. L., Hideg, I., & Spence, J. R., (2013). The culturally intelligent team: The impact of team cultural intelligence and cultural heterogeneity on team shared values. *J. Cross-Cult. Psychol.*, *44*(6), 941–962. doi: 10.1177/0022022113492894.
49. *Tight and Loose Cultures: A Conversation with Michele Gelfand* – Behavioral Scientist. https://behavioralscientist.org/tight-and-loose-cultures-a-conversation-with-michele-gelfand/ (accessed on 07 September 2023).
50. Cassell, (2019). Qualitative data analysis: Exploring themes, metaphors, and stories. *European Management Review*. Wiley Online Library. https://onlinelibrary.wiley.com/doi/10.1111/emre.12176 (accessed on 07 September 2023).
51. Rehg, M. T., Gundlach, M. J., & Grigorian, R., (2012). *Examining the Influence of Cross-Cultural Training on Cultural Intelligence and Specific Self-efficacy*. doi: 10.1108/13527601211219892.
52. Crowne, K. A., (2013). Cultural exposure, emotional intelligence, and cultural intelligence: An exploratory study. *Int. J. Cross Cult. Manag.*, *13*(1), 5–22. doi: 10.1177/1470595812452633.
53. Leung, K., & Morris, M. W., (2015). Values, schemas, and norms in the culture–behavior nexus: A situated dynamics framework. *J. Int. Bus. Stud.*, *46*(9), 1028–1050. doi: 10.1057/jibs.2014.66.
54. Haffejee, B., Young, C., & Corsun, D., (2017). The relationship between ethnocentrism and cultural intelligence. *Int. J. Intercult. Relat.*, 58. doi: 10.1016/j.ijintrel.2017.04.001.
55. Melissa, R. *5 Recession Lessons Hotels Should Remember in 2022 | Hospitality Net*. https://www.guestcentric.com/?utm_source=HospitalityNet&utm_medium=October2021&utm_campaign=HospitalityNet&utm_id=5RecessionLessonsForHotelsArticle (accessed on 07 September 2023).
56. Schawbel, D. *Brene Brown: How Vulnerability Can Make Our Lives Better*. Forbes. https://www.forbes.com/sites/danschawbel/2013/04/21/brene-brown-how-vulnerability-can-make-our-lives-better/ (accessed on 07 September 2023).
57. Nagda, B., (2006). Breaking Barriers, crossing borders, building bridges: Communication processes in intergroup dialogues. *J. Soc. Issues*, *62*. doi: 10.1111/j.1540-4560.2006.00473.x.
58. Raelin, J., (2011). From leadership-as-practice to leaderful practice. *Leadership*, *7*(2), 195–211. doi: 10.1177/1742715010394808.

59. Jiacheng, W., Lu, L., & Calabrese, F., (2010). A cognitive model of intra-organizational knowledge-sharing motivations in the view of cross-culture. *Int. J. Inf. Manag.*, *30*, 220–230. doi: 10.1016/j.ijinfomgt.2009.08.007.
60. Yershova, Y., DeJaeghere, J., & Mestenhauser, J., (2000). Thinking not a usual: Adding the intercultural perspective. *J. Stud. Int. Educ.*, *4*(1), 39–78. doi: 10.1177/102831530000400105.
61. Zhao, F., Deng, L., & Kemp, L. J., (2013). Interrelationships between cultural intelligence dimensions and the role of intrapersonal intelligence. *J. Gen. Manag.*, *38*(3), 3–24. doi: 10.1177/030630701303800302.
62. Korzilius, H., Bücker, J., & Beerlage, S., (2017). Multiculturalism and innovative work behavior: The mediating role of cultural intelligence. *Int. J. Intercult. Relat.*, *56*, 13–24. doi: 10.1016/j.ijintrel.2016.11.001.

CHAPTER 17

Artificial Intelligence and Cyber Security in the Service Industry

VINEETA KAPOOR

Freelancer Writer, Alumnus (2006) – Amity University, Noida, Uttar Pradesh, India

ABSTRACT

The methodology of identifying the scope, risks, and ways of handling artificial intelligence (AI) would differ almost in each sector. The analysis would take a few service sectors into account, recognize the problems that AI may pose, and explore possible solutions. This would include studying the reach of AI in Hospitality, Tourism, Healthcare, Banking, and Education verticals. Research on how the technology impacts or may impact each vertical positively and negatively would be conducted. The service industry focuses on customer satisfaction. While AI can have some beneficial benefits in the various industries of the service arena, it also has the potential of several risks attached. In order to negate or at least reduce these risks, specific parameters will have to be identified and implemented while putting AI into use. Ensuring that these smart systems for specific tasks will need to be in place, but a certain amount of control will have to be exercised so that the smartness of the system does not overpower human capability. This would include strategic decisions involving where and to what extent machines would be used to deliver service to customers. The most critical tasks would need to be addressed through a structured and well-defined approach. Identification of risk can be developed into an art and be used for AI as a concept.

The service industry creates intangible services. The purpose of the result is to understand the concept of AI and scrutinize its positive and negative effects in the chosen sectors of the service industry. To make the maximum use of AI, it is essential to identify its risks in each sector. The next step would be to tune it to the limit that no one takes undue advantage of the technology that conquers human sense and competence. A machine should remain a machine, however, controlled by a human. Its purpose is to reduce human labor, but only to an extent. It should not leave the skilled human resource redundant. Therefore, it is imperative to draw a line to define the limits of this smart system, to be disciplined by humans and not vice-versa. Further study needs to be conducted on the control system we need to build for the AI devices. For this purpose, organizations across various service industry sectors will have to plan a blended risk-specific control system to put a robust AI process in place. It will be incumbent for Protocols to be constituted and implemented for the ethics of AI to be applied to ensure a compelling customer experience.

17.1 ARTIFICIAL INTELLIGENCE (AI)

The duplication of human intelligence in machines to perform and conduct tasks like people is termed artificial intelligence (AI). This intelligent technology combines learning, rationalization, and sense. AI is continuously expanding to contribute to many industries. Algorithms are the basis for the development of AI [1]. The structure is such that simple applications are used for weak AI, and complex ones are derived for building solid AI. AI is a science of versatile nature with a range of approaches, but growth in ML and DL are generating a shift in pattern in virtually every sector.

17.1.1 WEAK AND STRONG ARTIFICIAL INTELLIGENCE (AI)

Alexa from Amazon, or Siri from Apple, are examples of weak AI. The system is designed to carry out one job. The main function that these devices perform is to answer queries. These systems cannot handle complex tasks. Strong AI devices are complicated systems that conduct tasks similar to humans [2]. These help resolve problems without human intervention. Such applications can be seen in hospitals banking, and many other areas.

17.2 IMPORTANCE OF ARTIFICIAL INTELLIGENCE (AI)

1. **Automated Discovery through Data and Repetitive Learning:** AI can handle continuous tasks without getting tired. It conducts computerized tasks of high volume and frequency.
2. **AI Puts Intelligence into Products in Use:** AI is normally not sold as an unaccompanied application. For instance, a new feature, Siri, was added to a new range of products from Apple. Thus, products already being used are improvised with AI-powered abilities. Large amounts of data can be added to mechanization, colloquial platforms for conversations, bots, and smart gadgets to enhance automation at home and the workplace, from security prudence to investment analysis.
3. **Adaptation through Progressive Learning Algorithms:** AI works based on algorithms so that the data do programming. AI enables the algorithms to find structure and regularities in data [3]. In the process, the algorithm becomes a predictor. Another AI technique is backpropagation, which allows the model to adjust when the first answer is not quite correct through training and additional data.
4. **Deeper Data Analysis:** This task can be accomplished using various layers of neural networks. This feature helps to create effective fraud detection systems using big data and computing power. Deep learning models require training since they learn from the data. These models become more accurate with more data being fed in them.
5. **Achievement of Incredible Accuracy:** Neural networks help in gaining absolute accuracy. Google Search, Google Photos, and Alexa are examples of deep neural networks. AI techniques are used in deep learning, image classification, and object recognition that can spot cancer on MRIs as correctly as eminently trained radiologists.
6. **Getting the Maximum from Data:** Data becomes precious when there are self-learning algorithms. AI needs to be applied to get the solutions out of the data. This makes data competent, and the

one that performs the best gets an edge over the others, among the many who use the same technique.

17.2.1 WORKING WITH AI

AI has not been created to take over human resources. It is a force to boost human abilities and make the work done better and quicker. The collaboration between human and AI have various benefits [4]. It increases the performance of technologies analytical in nature while removing barriers of language and translation along with economic boundaries. AI can also introduce newness to sectors where it is still not applied. The union also helps organizations to make work better with the present abilities. It helps provide a better understanding, better outlook and an updated perspective.

17.2.2 FUNCTIONING OF AI

There are various areas under the broad field of AI. It consists of a number of theories, technologies, and methods. The functioning of AI happens through a combination of huge amounts of data, where algorithms use different data patterns for quick processing:

1. **Machine Learning:** It is an area of AI which encompasses neural networks, operations, research, statistics, and analysis. It is an advanced way to find hidden information in the data.
2. **Neural Networks:** These are comprised of units that are connected to each other. A neural network processes data by reacting to external sources to share information between each unit [5]. This requires multiple data checks to connect and derive meaning from not defined data.
3. **Deep Learning:** This uses progressive computing power and better training methods for understanding complicated patterns of vast amounts of data. Image and speech recognition use the deep learning technique. It utilizes neural networks in large quantities to perform its tasks.
4. **Cognitive Computing:** Another wing of AI seeks a natural, human-like interaction with machines. It refers to a machine

designed using cognitive Computing and AI to replicate human processes by understanding speech and images and then responding accordingly.

5. **Computer Vision:** This perceives a video or picture, depending upon data patterns and deep learning. This aspect comes into use when machines process and understand images or videos.
6. **Natural Language Processing (NLP):** It is referred as the computer's ability to inspect, comprehend, and produce language and speech. The next level of NLP is interaction with computers for regular tasks.

   In addition, other technologies that support AI in various fields are graphical processing units, the Internet of Things (IoT), application programming interfaces, and advanced algorithms.

17.3 ADVANTAGES OF ARTIFICIAL INTELLIGENCE (AI)

The advantages of AI applications are many, practical enough to bring a surge in any industry:

1. **Reduces Human Error:** With the advent of AI, the complex tasks are undertaken by humans earlier, and which had great chances of committing errors, are taken up by technology to prevent the errors as far as possible. The machine applies algorithms used to collect information. For instance, AI used in weather forecasts has reduced maximum human errors.
2. **Risk-Taking Ability:** AI can be used to take up tasks involving vast amounts of risk, which would not be very convenient for humans [6]. The technology can be used in areas like mining, bomb diffusion, exploring Mars, etc.
3. **Round the Clock Availability:** AI can work continuously without any breaks. It does not experience tiredness or boredom. On the other hand, humans can work effectively for 6–8 hours, taking breaks in between. So, sectors where many queries keep pouring in, such as hospitals, educational setups, and helpline centers can employ AI to act in response to a higher consumer.

4. **New Inventions:** AI-powered software can help resolve complicated problems. For instance, in the medical field, predicting breast cancer in women using AI technology at early stages has helped save many lives.
5. **Digital Assistance:** Many organizations have customer support teams to handle the doubts and queries of customers. A voice bot or a chatbot can be set up to do so, to answer several inquiries, saving human labor and enabling them to focus on other job areas. Several companies have also deployed these bots on websites and mobile applications to chat with customers.
6. **Making Repetitive Jobs Convenient:** Many tasks need to be done in a repeated manner for each customer. AI can handle such tasks for quick turnaround and minimum chances of errors. For example, banks have the same document verification process for loans every time. AI can undertake this work. The time consumed would thus be reduced to a great extent with error-free results, thus benefitting the bank and the customer.
7. **Quicker Decisions:** The programming is made in a program enabled machine works purely on the algorithm fed in it. It does not analyze the pros and cons, unlike humans, and therefore derives results instantly. The chess game in Windows uses algorithms that make the game win every time!
8. **Daily Applications:** AI has been embedded in our daily without us realizing it. Siri, Cortana, Google Maps, voice-based recognition are all examples of AI we use in everyday actions—going somewhere? Ask OK Google, and it will show you the location and other details at the snap of a finger!

17.4 DISADVANTAGES OF ARTIFICIAL INTELLIGENCE (AI)

Like everything else, AI also has some disadvantages, some of them can be explained as follows:

1. **Unemployment:** One of the biggest threats humankind fears is that AI can replace human force in organizations to have more efficient results in work. The less qualified employees are at the

highest risk of being turned down once this technology is implemented full scale.

2. **Expensive Production:** AI requires vast amounts of data to be fed into it as algorithms to produce effective results. This data cannot be scattered because it has to be furnished for algorithms to use it. This involves cost. Further, the machines require continuous maintenance and repair that is expensive to carry out all the time.
3. **Making Humans Lethargic:** Human labor becomes relatively minimal, with most of the tasks being assigned to AI-based machines and software [6]. Unless people engage themselves in other productive work, they become lazy and start depending heavily upon technology. This might cause a severe problem in health and other related aspects in future.
4. **Emotionless Technology:** There are many areas where human touch, care, and understanding become essential to make things work. In a classroom, a teacher can well understand what a particular child might be going through if he/she might be unable to focus on class. A caring nurse in a hospital cannot be replaced by a bot just providing medicines and food on time.
5. **Lacking Newness:** Machines can perform tasks only according to their instructions. They are incapable of thinking something innovative. Also, if they crash, the data could be lost along with the work. Humans can always recollect and create new ideas, even if work faces a setback. However, the drawback with machines is that they cannot think of avenues on their own to make things work.

17.5 IMPACT OF AI ON GOVERNMENTS

The governments of various countries can incorporate AI in different spheres to improve their functioning. One such area is public security and safety. Biometric facial recognition technology is coming into use in several government offices to provide security to the people and information in these setups. It can also help in border control and identifying criminals [6]. AI-based ML can be utilized for the prediction of crime and may assist to prevent it. Further, efficiency at the bureaucratic level can be attained by adopting certain AI-powered techniques. The government personnel can save time by assigning repetitive tasks to the machines. The technology also helps in reducing problems like red-tapism, thus enhancing the efficiency and effectiveness of government systems.

17.5.1 HOW IS ARTIFICIAL INTELLIGENCE (AI) IMPACTING THE SERVICE INDUSTRY?

Deployment of AI technology within the industry is being noticed in many spheres and levels of labor. To put a sound AI system in place, stakeholder groups should actively get involved in determining a way to apply AI in the development of services to serve the shoppers better [7]. The industry always aims for customer satisfaction. So, to benefit the customer and society, AI is becoming an indispensable tool to resort to. Therefore, initiatives are being taken to make this technology a part of the system across most service sectors.

17.5.2 ARTIFICIAL INTELLIGENCE (AI) AND CYBER SECURITY

AI is becoming a compulsory part of cyber security, given the cyber threats that organizations across the service and other industries face. The resource limitations, technical glitches, and risk of cyber security have to be addressed, and AI seems to be the only resort. To accomplish this, the key stakeholders and prominent corporate representatives of the industry must collaborate within their respective sectors to identify the appropriate use of AI. The security system's structural rigidity stipulation must be ascertained, to build AI-powered security accordingly. Organizations should devise new methods of audit using AI technology to prevent any breach of data [8]. They should also enhance AI productivity and reliability. Handling the data libraries becomes the foremost important task and has to be done with maximum alertness. All these software libraries should be lined with AI codes that might help maintain the absolute security of data. The staff of the organizations must be trained enough to acquire skills to manage the AI processes and cyber security to own a balance between them. Let us examine the application of AI in a few selected sectors of the enormous service industry.

17.6 ARTIFICIAL INTELLIGENCE (AI) IN BANKING

The Banking Industry is a network of economic institutions which offer a gamut of financial and banking services. Banks treat various sorts of wealth to store, transfer, extend credit, or manage risks associated with it. In any economy, banks are the main drivers and supply the liquidity needed for families and businesses to invest in various avenues for the future [9]. People take credit or bank a loan that helps the families be able to buy a house or take up education without having to save all the time. Many organizations can start new businesses or expand with the assistance of loans. Banks can utilize AI that can innovate methods to reach out to people. There is a requirement for pace that necessitates the use of humans and AI to steer deployable and procedure efficiencies. While implementing AI, the consumer and staff satisfaction can be advantageous to growth. The AI in Financial Services global report found that 85% of respondents already use AI in some capacity for convenience and accuracy, and 77% of respondents said AI will be one of their key investment areas in the future [10].

17.6.1 USES OF AI IN BANKING

Front-office and middle-office applications of AI can significantly save costs. In order to have proper authentication of the customer, AI is being adopted for front-end operations in banks. This is done through chatbots and other virtual assistants for an interesting, accurate interaction with the customer [11]. It also helps in strengthening relationships with customers and recommending the proper thing. Banks deploy AI within the middle office for fraud protection and anti-money laundering processes. The know-your-customer (KYC) checks can also be tracked more efficiently in this manner. Banks are progressively executing plans to integrate AI throughout the whole banking sector [10]. The technology may well be used to achieve partnerships with external businesses, hire appropriate employees, reuse data and reach out for more business.

17.6.2 POPULAR APPLICATIONS OF AI IN BANKING

The below are a few applications of AI that we see in the banking segment:

1. **AI Chatbots:** Chatbots are used extensively in banks to interact with customers and deliver them with a personalized feel. These services are used to promote the brand within the market.
2. **Seamless Process:** Reduction of workload for the workers, efficient response, accurate information, and around-the-clock availability – these are the advantages of automation in AI in the banking industry.
3. **Portfolio and Wealth Management:** The AI technology enables people to manage the banking services and the mobile applications for better wealth management [12]. AI additionally ensures automation of the debit and credit card management system. One gets a message to test for any fraudulent activity on making an internet transaction or suspecting an activity: this and other AI features help make transactions safer.
4. **Risk Management:** While disbursing loans, it becomes essential for banks to assess the financial activities of the borrower and then analyze the risks attached, if any, while giving loans. AI comes into play here because it can easily track financial activities through mobile applications and thus reduce fraudulent act cases.
5. **Enhanced Customer Experience:** According to the searches made by customers, AI-based mobile banking applications offers services and schemes [13].
6. **Data Collection, Analysis:** AI technology uses authentic data sets to assess the information and effectively perform data collection and analysis. AI-powered mobile banking apps collect and create an appropriate learning procedure for enhancing the general user experience. Personalization of data can be done in a much better manner after data analysis.
7. **Analyzing Market Sentiment:** Machine learning helps AI assess market conditions and determine market trends. It then becomes easier for investors to make financial decisions. It also aids in fund allocation for various functions.

8. **Next-Gen Security:** AI ensures safe and secure transactions with banks. Expert minds work behind the event of AI banking applications to produce high-level security for banks and customers.
9. **Fraud Detection:** AI banking applications can track fraudulent acts or irregular banking patterns, thus minimizing the possibilities of fraud [14]. Banks also can protect themselves from their security being breached with the assistance of AI.
10. **Compliance Issue in Banking:** AI technology is spreading out very swiftly. The largest organizations examine banks that follow regulatory compliance rules. The update process in banks is tedious and takes much time when done manually. The AI system enables quick and continuous updation, resulting in inaccuracy in work.

17.6.3 AI DISRUPTING THE BANKING SYSTEM

The small start to disrupt the banking system hugely has already begun. The only prominent kind of AI in this sector is the chatbot. An AI program called a digital voice assistant enables users to communicate with a machine. With the aid of deep learning algorithms and a rule-based approach, chatbots can converse with people in natural language. Chatbots in banking provide an interactive experience while answering common queries that customers wouldn't need to ask a human about [14]. The consumers get a contented digital experience, while banks can lower their prices.

The AI chatbot from the Bank of America, Erica, can handle card security updates and reduce credit card debt. During 2019, Erica handled over 50 million client queries [15]. Chatbots are created to support and facilitate tasks for banks that consume plenty of time. At no point are they meant to be exchanged for the human workforce. The Indian Banking Space has also come a long way with technology and is well-equipped to serve tech-savvy customers with online services, ATMs, eKYCs, and many other features. The State Bank of India has the SIA, an AI-enabled virtual assistant, handling 10,000 banking enquiries per second. Eva, introduced by HDFC, aids in providing information on banking products and services to customers. Apart from chatbots, many banks are using robotics software to ensure smooth back-end processes for better functioning of these organizations.

17.6.4 HUMAN WORKFORCE AND AI

AI should aim at reducing the burden of humans by investigating the data, which takes a lot of time if done by humans. At no point, technology should replace human resources. There must be a collaboration and sustainable relationship between the machine and humans, giving rise to a hybrid workforce. Cyber security is an area to be focused upon, where both the resources can work together for amazing results. Further, it is required that the banking institutions produce their in-house machines and train their human workforce to handle the systems. There would be new roles in the team: cyber security experts, data scientists, data analysts, network engineers, and IT professionals with diverse skills.

17.6.5 ISSUES WITH AI

According to World Economic Forum, data will become crucial as a differentiation point for banks. Therefore, consideration for data monetization through Big Data analytics would be critical. For this, banks need to have structured data to allow machine learning models to be trained effectively. Machine learning would be the next step once big data and data mining are mastered.

17.6.6 AI BLACK BOX

The finance sector is constantly under scrutiny, making it imperative to have a robust AI model. One issue with AI is that we can see the decision made by the algorithm, but we cannot see the working of the same. This is called the AI Black Box. AI systems have to find a degree of transparency in operations like banking to avoid any loopholes.

17.6.7 LEGAL LIABILITY

The banking sector must consider AI's legal and ethical liability. As per the data protection rules, customers should be aware of how their data is being utilized. There is a need to maintain a balance between customer interest and innovation. Thus, the regulatory function should be free of any bias.

17.6.8 TESTING OF AI

AI must be in a safe environment. In a sector like banking, it is vital that the system is foolproof before being implemented. There are financial exchanges and deals in this industry, so there needs to be a robust AI to handle all the queries accurately.

17.6.9 CHALLENGES

One of the main challenges faced by the banking industry while adopting AI methods is the digitization of data. All banks do not have data on computers and are using papers to store the same. Implementation of AI requires an infrastructure with a digital framework [17]. There is a cost involved in adapting to AI technology. However, the integration of this smart technology with the human workforce can be an effective and highly useful one if collaborated with proper planning.

17.6.10 THE FUTURE WITH AI

The customers want a hassle-free experience at the banks and AI can assist. They would prefer services conducted through the banking apps if given a choice. AI has become the present of the banking sector, not the future. The industry is experiencing a revolution with more data and newer technology [1]. People want smooth experiences and fantastic results from every service they use. They can resort to other fintech companies if banks cannot satisfy their needs. Keeping this in mind, banks would have to corroborate data, people, and processes to work effectively.

17.6.11 SUMMARY – AI IN BANKING

AI is improving the banking industry in several ways. It provides better customer service with the introduction of bots in many financial organizations, which can easily handle customer queries. Robot financial advisors are coming into the market, which hardly makes any errors and offer robust financial advice and portfolio management services. They provide less paperwork, greater efficiency, and better financial security in money laundering and fraud. Thus, the significant AI-powered features are gradually being recognized as beneficial.

17.7 ARTIFICIAL INTELLIGENCE (AI) IN EDUCATION

The Education industry, a multi-billion-dollar industry, consists of entities that provide education. The education techniques for instructing, training, teaching, research, storytelling, and discussions. This sector includes academic institutions and many other service providers to facilitate learning. With the advent of technology, education has become broader and the future brighter. Education is not just restricted to classrooms now. It is beyond that. The integration of modern methods like AI has paved a path for upward growth in the education sector, opening innumerable avenues. AI techniques like machine learning, NLP, and neural networks are being applied in various areas of education to make learning innovative and more enjoyable.

17.7.1 ARTIFICIAL INTELLIGENCE (AI) IN THE GLOBAL MARKET

It is predicted that the world AI in the market for education will reach USD 3.68 billion in value by 2023, growing at a CAGR of 47% between 2018 and 2023. This growth would also increase the demand for multilingual translators integrated with AI technology [18]. Given the growing desire for AI, the education sector is expected to see enormous growth in AI.

17.7.2 NATURAL LANGUAGE PROCESSING (NLP) TECHNOLOGY – PREDICTED TO GROW AT A BETTER CAGR

This is when compared to deep learning and machine learning. The expansion is principally because of the rise in virtual classes in K-12 schools and universities, resulting in technology investments for effective imparting education [19]. Further, NLP is being used to understand different human languages and read the content stored in digital formats in different data sets.

17.7.3 FUTURE GROWTH OF SIMULATED IMPLEMENTERS AND LEARNING ENVIRONMENTS APPLICATION

This includes the organization of students academically, administrative coordination, use of digital technology in education, and real-time communication. The AI-enabled virtual facilitator supports the instructor and acts as a bilingual interpreter and student mentor. These AI-enabled educators and surroundings accommodate students' learning needs [20]. Moreover, students using virtual environments can solve real-time social problems associated with social development. It also enhances their knowledge and widens their viewpoint.

17.7.4 RECENT DEVELOPMENTS

IBM joined hands with Salesforce in March 2017 to offer IBM Watson and Salesforce Einstein services, thus attracting more customers. With this partnership, both companies' combined Watson's capabilities with Einstein to rope in many customers. Amazon entered a collaboration with Adhesive, a computing curriculum provider in schools, in 2018. According to the association, Amazon web services (AWS) would subsidize schools to apply Edhesive's curriculum. In March 2018, Microsoft developed Cognitive Services tools such as Face API, Microsoft Custom Vision Service and Bing Entity Search. Developers could Quickly add AI capabilities to various applications with new cloud-hosted APIs. In February 2018, Google launched an AI website, learn with Google, to deliver and encourage training in AI. At the same time, Pearson associated with Microsoft Research Asia (MSRA) for three years for integrating AI logic into the market-leading English language learning curriculum. Drones are already being employed to test possible applications to monitor whether social distancing guidelines are being followed during COVID situations by several jurisdictions, including the US [21]. More advanced applications are on the verge of being implemented, like drones with the potential to detect COVID-19 symptoms, like identifying individuals with high temperatures in a crowd.

17.7.5 OPPORTUNITIES WITH AI IN THE EDUCATION SECTOR

AI technology can analyze data to assess the conduct of scholars. The smart system eases the processes of educational institutions and improvises relations between them and other organizations. The academic institutions have their AI platforms on which the AI tools are developed. Such tools help students resolve academic problems and requirements, thereby enabling them to increase their productivity [22]. All these factors indicate the need for computing techniques to reinforce the chance of the growth of AI in the education sector.

17.7.6 CHALLENGES FACED BY AI IN THE EDUCATION SECTOR

The AI-powered assessment software is not equipped to assess the answers like true/false, fill in the blanks, short answers, essays, and multiple-choice questions. These answers require logical thinking, circumstantial prudence, and content veracity. Upscaling the level of AI is required to a great extent to cater to these requirements and be appropriately used. How AI can help in education, following are the few examples where it is already impacting:

1. **Communication:** AI has the flexibility to connect students and teachers and instantly begin with the communication process from anywhere in the world. It would be highly convenient to connect with peers worldwide and make friends virtually. A student's needs and error analysis are done, and peers are recommended. AI also provides countless learning possibilities, such as living in another country or building foreign language skills.
2. **Differentiation:** The resources required by students and teachers on the internet can be quickly sorted and presented before them, rather than they are searching for the entire web [22]. A student can also have immediate access to a tutor whenever needed with the presence of AI.
3. **Personalization:** AI can analyze student responses, interest areas, and provide them with more learning opportunities to understand the course content better. AI exhibits the quality of letting the teacher know about the virtues and shortcomings of each pupil, hence assisting the teacher give personalized attention to each student according to their attributes.
4. **Exploration:** Students can benefit more in the classroom by introducing virtual and AR. These tools would help them see the

place they would otherwise be unable to explore newer things [23]. Through AI, student responses can help search for resources, which can further become an experience for the entire classroom. AI can bring the content to life and show what the students want to see.

5. **Assessments:** AI can help teachers mark student performance and streamline the assessment process, with the added advantage of quickly taking the data, providing analysis for teachers, and having ample time for classroom teaching. The faster results that AI can achieve allow each student to learn and grow.

17.8 PROS OF AI (EDUCATION PERSPECTIVE)

1. **Virtual Reality (VR):** This feature allows educators and students to venture into 3D virtual world experiences, helping them expand their knowledge immensely.
2. **Time Efficiency:** Teachers can focus on more things when AI can handle repetitive tasks like attendance.
3. **Personalization:** A teacher can assess each student's personality and learning pace by using customized AI programs that include students' data.
4. **24-Hour Access:** AI becomes very useful when it can be utilized at any hour of the day, making it flexible for students to access it at their convenience [24].
5. **Reduces Human Error:** Since AI is based on algorithms, the chances of human error become nil.
6. **Expanded Outreach:** AI enables students to attend classes from anywhere if they cannot come physically.
7. **Closes Language Gap:** AI enables real-time translation, and this aids in improving interaction in a class with no language barrier.

17.9 CONS OF AI (EDUCATION PERSPECTIVE)

1. **Experience No Consideration:** No experience AI gains over time. It cannot become better with time in its answers or knowledge as it is algorithm-based.

2. **Removes Classroom Culture:** A traditional classroom does not allow the use of technology. With AI, technology becomes compulsory to use, leading to its addiction.
3. **Reduces Personal Interaction:** The relationship between a teacher and a student would gradually perish if AI ruled the classroom [24].
4. **Unemployment:** This would be a growing concern, as teachers would not be required if machines do all the work.
5. **No Flexibility:** As AI is purely based on algorithms, there is no flexibility in teaching methods that can be provided as per the needs of students.
6. **Reduction in Hands-on-Skills:** Society is bound to suffer as human skills without technology would not be sharp enough, with technology making things easier for us.

17.9.1 CHALLENGES FACED BY TEACHERS – ROLE AND POSITIONING

Education aims to promote spiritual growth, dig out potential and recognize the capability of people. Teachers would be required to do their bit at any age, along with changes in the ways education would be imparted. Teachers can understand students' physical and mental aspects, which AI cannot. Teachers need to be open to innovation and learning, but the technology should be used only to adopt the best teaching methods [25]. The tech machines cannot replicate the caring ways of teachers. Understanding the problems of students and communicating emotionally with them can only be done by humans. Thus, the role of teachers might become easier but will not diminish.

17.9.2 FUTURE OF AI IN EDUCATION

AI can quickly interpret a student's needs and design an appropriate assessment. It can show students' expertise in a field or subject and create an individualized plan for them. AI also lets the teacher have more time to concentrate on class needs and facilitate learning by providing a virtual assistant [26]. The technology can also benefit parents by keeping them informed about their children and helping get notes if sometimes the classes have not been attended by students. There are avenues where parents, teachers, and students can interact openly and effectively.

17.9.3 SUMMARY – AI IN EDUCATION

With AI's potential, research into a new arena of technology in the field of education must be conducted. The educators must revisit and redesign the curriculum for all levels of students, which in turn would be used by AI to build applications. Educators and researchers should continuously upgrade themselves in new applications in AI to guide students in making an ethical decision about the use of such systems.

AI and machine learning can occupy a unique position in all educational experiences [27]. The future of AI looks bright in the education industry is expected. Though the future of AI in the education sector is not yet furnished, it is clear that students with various capabilities would have better chances of learning and moving towards growth. AI can help students analyze and choose their careers, the education required to achieve, and determine their career goals. Educational institutions can build more innovative next generations by investing in AI.

17.10 ARTIFICIAL INTELLIGENCE (AI) IN HEALTHCARE

The Healthcare industry includes hospitals, clinical trials, telemedicine, medical tourism, health insurance, medical equipment and outsourcing. By creating intricate algorithms to carry out tasks automatically, AI is being used in the healthcare industry. The doctors, scientists, and researchers feed in data into computers, and new algorithms are formed to review, decipher, and even suggest unraveling complicated medical problems. The discussion of the relevance of healthcare with AI are given in subsections.

17.10.1 MACHINE LEARNING AND NEURAL NETWORKS

Precision medicine employs machine learning to anticipate which treatment regimens will be effective for a patient considering the recommended course of action. This aspect requires training datasets to reach a known conclusion and happen under supervised learning. Another machine learning method is the neural network, which is used to analyze issues in terms of inputs, outputs, and the weights of the parameters that link inputs and outputs.

17.10.2 NATURAL LANGUAGE PROCESSING (NLP)

Natural language processing (NLP) comprises designing, understanding, and arranging clinical documentation and published research [28]. NLP systems can examine disorderly clinical notes on patients, report preparation, interpret interactions with patients and conduct conversational AI.

17.10.3 RULE-BASED EXPERT SYSTEMS

These systems require the proficiency of humans and knowledge engineers to build a rule in a specific field. However, these systems are challenging to implement if the number of rules is enormous and difficult to implement. They start breaking down. Such rules are gradually being replaced by machine algorithms now.

17.10.4 PHYSICAL ROBOTS

It would not be surprising to know that 200,000 plus manufacturing robots are established each year around the globe. Jobs like delivering supplies in hospitals, lifting, shifting, welding or assembling objects in factories and warehouses are done by them. AI features fed in some of these robots make them smarter and intelligent, capable of following human instructions. Their operating systems are their brains where the AI capabilities are being fed [29].

17.10.5 ROBOTIC PROCESS AUTOMATION (RPA)

Administrative digital tasks which involve information systems are being carried out by this technology. It is similar to a human abiding by a script or rules. RPA systems work on computer programs and do not comprise of robots. Though they are expensive, they are easy to be programmed and implemented.

17.10.6 DIAGNOSIS AND TREATMENT APPLICATIONS

One of the main focuses of AI has been diagnosing disease and then treating the same. For the same reason, MYCIN was developed at Stanford to identify diseases caused by blood-borne germs. Unfortunately, it did not prove to be very effective compared to human diagnosis capabilities. Other technological applications of AI include medical diagnostics for diseases, the discovery of drugs by many health and pharmaceutical companies, and clinical trials being redesigned [30]. Patient outcomes can be improved through various strategies motivated by AI. Another upcoming area is pain management, where the patient can be diverted from pain and helped.

17.11 CHALLENGES AI INTEGRATION IN HEALTH SYSTEMS

17.11.1 INTEGRATION OF AI INTO CURRENT TOOLS

Workplace workflows and business intelligence (BI) solutions must be AI-powered. The hindrance faced here is that most AI tools require another program for the analysts to learn and implement. This can be a limitation while adapting to the AI methods.

17.11.2 INTEGRATING AI EXPERTISE WITHIN THE ORGANIZATION

The implementation of AI in the organization requires data experts and others to fill the gap. Moreover, this can be challenging for myriad reasons:

- Health care systems may find it challenging to finance data resources.
- In-house data engineers, proficient analytics teams are tough [31].
- Teaching new skills to the teams working on other systems is costly. It slows down the process and could be unsustainable.

17.11.3 THE IMPACT OF AI (POSITIVE)

Healthcare businesses struggle to show the benefit and build trust across the organization even after they have successfully implemented predictive models. The most typical difficulties include:

i. **Limitations of "Black Box" Software:** Most programs display the response or result and do not display analysis process, leaving the stakeholders to wonder how these programs function.

ii. **The Narrow Focus on Predictive Analytics:** Data analysts use predictive analytics to predict the outcome by testing data and records. But the focus on this area is not much, which hinders the potential for a greater impact in this field.

17.11.4 DEVELOPMENT OF NEW MEDICINES WITH AI

A robot named Adam was put to use to research yeast in 2007. Another robot, Eve, was involved in discovering a common ingredient in toothpaste, called triclosan, and could fight malaria-based parasites.

Innovations in the drug development industry require considerable costs to be incurred, and thousands of person-hours are put to use [32]. Putting each drug to clinical trials is around $2.6 billion. With so much hard work and expenditure, just about 10% of medicines enter the market. To avoid these unnecessary expenses and labor, the medical fraternity is looking towards AI technology to take over in experimentation and analysis to have accurate results involving less expenditure and saved human labor.

17.11.5 MINING DATA WITH AI

AI technology is being looked at by many in the healthcare industry as it has several advantages [33]. Data cannot be lost or damaged while being used by AI-powered software. The intelligent system can break down and bifurcate data in minutes, which used to take years. Further, the ability to connect data points is possessed by AI. Thus, the healthcare sector is starting to rely heavily upon technology.

17.11.6 AI ROBOT-ASSISTED SURGERY

At the famous Mayo Clinic, robots assist doctors in conducting complicated surgeries with utmost precision, proving that robots can handle crucial tasks in the healthcare sector. Many other hospitals have started using robots in surgeries, even as critical as open-heart surgery. Several doctors have started believing that these machines can handle such critical tasks with meticulousness and resilience that is sometimes difficult to find in humans. Robots have cameras installed and are supplied with mechanical arms and surgical instruments to further the experience. Along with skilled doctors, they even create new kinds of surgeries [28]. Surgeons control the tech machines while seated in front of computers, and the robots provide them a multi-dimensional, augmented vision of the precise operating site that surgeons might not get with bare eyes. The surgeon and the teamwork with the robot to conduct the surgery in this manner. It has been noticed that surgeries assisted by robots have proved to have fewer complications, quicker recovery time with less pain. Robotic surgery has reduced the complexity of surgery, reduced pain, and shortened recuperation times.

17.12 USEFULNESS OF AI IN HEALTHCARE

AI is used in a number of fields in the healthcare sector today.

17.12.1 RADIOLOGY

A radiologist can use AI to increase efficiency while automating image analysis and diagnosis. It helps in reducing human mistakes. The technology can automatically analyze a scan and does not require human assistance. An organization in the USA has even sought Government approval to use an AI-powered platform for analyzing and interpreting cardiac MRI images.

17.12.2 PRIMARY CARE

Several organizations resort to AI methods and advise on a voice or chat-based interaction. This provides quick and accurate access to basic queries and medical issues. This could help provide basic guidance to people even in remote or under-served areas. However, such solutions are still under research to ensure patient safety and data privacy.

17.12.3 DRUG DISCOVERY

The extensive databases of information on the existing medicines are being used by AI solutions to develop and identify new potential therapies [34]. This step can help enhance the effectiveness and success rate of drug development, thus encouraging the process of bringing new drugs into the market to combat life-threatening diseases.

17.12.4 IDENTIFYING PATIENT RISK

To identify patients at risk, the clinics can use AI solutions to analyze large amounts of data of the old patients. It needs to be ensured that once the treatment of a patient is complete, there is no chance of re-admitting the same patient for the same problem within the next 30 days. More work comprises of the capability of analyzing the still image of a patient's retina and anticipating the risk of cardiovascular disease.

17.12.5 POTENTIAL RISKS OF AI IN HEALTHCARE

A vast amount of patient data is required to incorporate AI solutions to train and optimize the performance of algorithms. However, the process can pose a range of risks:

1. **Quality and Usability of Data:** Data can be pretty subjective compared to other industries. There could be issues like unsystematic notes maintained by the clinician, which could be challenging to work on. Further, the data could also be inaccurate, and details of patients could have been entered wrongly or misleadingly.
2. **Patient Privacy:** A very crucial point, the personal medical records of patients need to be protected so that no misuse takes place with the information [34]. Seling such information to AI companies is another question that is yet to be pondered upon and depends on a

case-to-case basis. In Europe, the General Data Protection Regulation (GDPR) allows people to get their data erased if they do not want to keep records anywhere. Another point to be considered is that if AI companies utilize data for various purposes, including analysis of disease, will the company be liable to ensure that the patient is not in danger?

3. **User Adoption:** The medical sector has the impression of doctors and nurses to be around for check-ups, suggestions, and care. It is to be seen that in this field, where we still use fax machines, will AI technology be accepted when it comes to a machine carrying the entire diagnosis and a software algorithm determining the illness?

17.12.6 THE FUTURE OF AI IN HEALTHCARE

Hybrid models would provide the healthcare industry with immense opportunities to receive diagnosis support, identify risk factors, and plan the solution. However, the responsibility of the patient's wellbeing would be the top priority [35]. Once the above is assured, it would be convenient for healthcare providers to adopt the technology faster and easier. It would, thus, help deliver patient outcomes and operational efficiency at scale, which is significantly enhanced after mitigating perceived risk.

17.12.7 SUMMARY – AI AND HEALTHCARE

It is imperative to have a robust technology in place to resolve the wide range of problems in the healthcare industry. AI is the way to find innovative solutions to the cropping issues in this sector. The impact of AI-powered solutions is yet to be made, and the technology can resolve a large number of complications involved with various diseases. These complications and challenges need to be addressed in the coming times now, and AI seems like the only resort for them in the global healthcare industry.

17.13 ARTIFICIAL INTELLIGENCE (AI) IN TOURISM

Artificial intelligence (AI) technology can revolutionize the entire tourism sector. Though it has not been utilized in the travel and tourism industry to its fullest potential, AI can ensure customer satisfaction while redesigning itself to the changing ways of the industry. Customer service could have a different way with AI. The system can take up helping to identify travelers while they are traveling and customize a person's travel needs. It can also function on its own even without the staff of airlines and other related segments.

17.13.1 BIG DATA

Big data comprises an array of data with huge volumes developed at high speeds from varied sources and multiple sources of data. This concept revolves mainly around four Vs: volume, velocity, variety, and veracity. In tourism, Big Data is derived from environment and tourist sources. The environment is the origin of events happening at a destination, IoT, transactions, information gathered from sensors and meteorological data [36]. The second source for Big Data, tourists, provides data differently. They search for travel and accommodation information before traveling and then make bookings for the same through which their data can be captured. During their trip, their travel leaves traces of their movements, encapsulated by data from mobile, GPS, Bluetooth, etc. Biometric and emotional data can also be gathered similarly during the trip. Lastly, user-generated content (UGC) is collected when tourists post pictures or videos or send online reviews. All this content is taken together to create a user profile, which is used to suggest products and services to the users.

17.13.2 MACHINE AND DEEP LEARNING

Machine learning is carried out with huge data sets. There are a set of algorithms used by the machine, where they learn processes, repeat them and derive conclusions. These conclusions can be fed by humans or derived by the machine itself. On the other hand, Deep Learning uses small sets of instructions to evaluate and modify a model to make it effective and more accurate [36]. These concepts predict the weather, sentiment analysis, translation, forecasting, image and video recognition, and fraud prevention.

17.13.3 NATURAL LANGUAGE PROCESSING (NLP)

Natural language processing (NLP) permits computers to deal with natural language properly. This process involves input which can be through voice or text. In the case of voice input, a voice recognition process is carried out before the language processing. Natural language generation enables conversation with the end-user through the IT system and is a part of NLP [36]. NLP is one of the requirements of automated translation. NLP facilitates virtual assistants, conversational apps and robots, and thus its relevance in the tourism sector is valuable. There are various areas where AI can be brought into use effectively and can boost operations in the tourism industry, which are discussed in succeeding sections.

17.13.4 INCREASED CUSTOMER SATISFACTION

Chatbots are AI programs designed to interact with customers through chat interfaces. They can answer queries and even suggest suitable options. These programs can considerably enhance the customer service experience [37]. AI chatbots can be customized to support customer loyalty programs and assist customers. Chatbots help streamline things and provide prompt answers to the customers. Another feature that allows for more accessible travel formalities is facial recognition. It aids in a quick process when it comes to proofs and identification.

17.13.5 DATA – PROCESSING AND ANALYSIS

AI has the competence of helping to design business plans and formulate pricing strategies. It can gather data and analyze the same to reach conclusions in different matters. Data can also be sorted quickly and efficiently. AI can also benefit organizations by customizing travel packages after creating profiles of customers and foreseeing their travel needs. The travel industry can be the leading industry using AI and its gadgets effectively.

17.13.6 VIRTUAL REALITY (VR) AND AUGMENTED REALITY (AR)

Virtual reality (VR) is a growing digital environment that allows the customer to experience the product or service they would consume in the future. In the travel industry, VR enables one to glimpse what would be experienced in reality. For example, booking procedures are made easier with this technology in place by giving the customer a complete booking process experience. A person can seek all information required, from flight details to hotel options to local sightseeing. A booking can be made simultaneously without using a computer mouse or a touch screen [38]. Further, there are virtual hotel tours for customers to know what their hotel and room would look like or if they can look for an alternative. The travel sector can sell its products and services by giving consumers a prior experience through VR. How would it feel to be at the majestic Opera House in Sydney or standing at the top of the Eiffel Tower in Paris? VR provides a complete experience package within the comfort of one's room [39]. AR can be operated through smartphones, tablets, and other devices. It is an excellent tool in marketing, used to add graphics or relevant information through a compatible device. Talking about tourist destinations, a person can augment various locations and know more about particular areas or restaurants by pointing at them. AR is used as a tourist information tool in some hotels, where the guest can access wall maps in their hotel rooms and find extra information about the sights to be visited. Innovative thoughts have been put into this tool to change the way customers perceive their travel surroundings, making it very useful for businesses and marketers.

17.13.7 ONLINE REPUTATION MANAGEMENT

With the advent of AI, organizations can keep track of their customers concerning their experiences, grievances, and feedback. Online communication in this manner can happen more quickly and effectively.

17.13.8 ENHANCED PRODUCTIVITY

AI helps in reducing the scope of human error. It also enables quick completion of tasks, along with handling multiple jobs at the same time. This would finally enhance productivity to a great extent and benefit the organizations. For example, Mildred, the bot by Lufthansa, helps customers find the cheapest flights. KLM Royal Dutch Airlines also utilizes AI technology to answer queries on Twitter, Facebook, etc. Face-to-face customer interactions are key to customer satisfaction in the tourism sector. Being a service-based industry, interaction becomes primarily essential [40]. Keeping this in mind, AI has been deploying technologies for queries to be answered in no time, without the customer having to stand in long queues, leading to an efficient system. Airlines like KLM started handling 50% of their inquiries through AI by the end of 2017. Hotel operator 'Dorchester Collections' adopted the technology and used customer feedback and reviews to customize its menu options for a better customer experience. The iPhone app, Lola, helps book flights and hotels.

17.13.9 SOCIAL MEDIA

There are many social media listening tools, and some are particularly designed for the travel industry. Customers can use these applications to post their thoughts, feedback, suggestions, grievances, and even their experiences at different destinations or with a tourism operator. These can, in turn, be helpful to tourism organizations to customize products according to customer preferences and provide a better experience.

17.14 CHALLENGES FACED BY AI IN THE TOURISM INDUSTRY

Apart from the several uses and benefits that AI brings with it, there are certain challenges that the technology faces in the tourism sector:

- While AI can help tourists have prior experience of unknown destinations, they feel there is a threat to privacy as it gathers data for patterns and information of travelers. The more significant concern is the dominance of technology in society, and tourists would have to choose between fully automated, low-cost services and less mechanized, human-based services.
- Another implicit constraint is the substitution of human beings with AI [41]. The growing AI technology is capable of replacing humans to a large extent. The front desk could see robots handling customer queries and all functions. The personal touch provided by humans could well be lost with the introduction of robots and similar machines in various tourism areas.
- AI is an able technology to adapt to any feature fed into it. A concerning area of bias that humans have practiced for ages would have to be carefully handled while dealing with machines. It would be essential that ethics be a part of the AI systems so that unbiased decisions are made and all participants are benefitted.

17.14.1 WHAT NEXT?

Having analyzed the benefits, challenges, and concerns with the AI systems, conducting complete throttle research while developing the system becomes imperative. There are some areas to be kept in mind to consider for a robust technology. The sustainability of a tourist destination is a prime concern. The impact of AI on destinations has to be evaluated and managed accordingly. Secondly, research on data collection methods from tourist activities to better understand their preferences needs to be monitored to avoid privacy breaches [42]. AI would also affect the working patterns in the sector in various areas. These areas have to be identified, and appropriate usage must be assured. Lastly, the impressions of AI on tourism marketing and operation activities have to be analyzed for effective use of the technology in such work areas.

17.14.2 SUMMARY – AI IN THE TOURISM INDUSTRY

The AI applications reduce the time taken to accomplish work while improving the correctness of procedures and results. These are precious facilities in an industry where time is essence and information is constantly changing. We are now at the cusp of the third phase of (AI) in the tourism industry. AI is becoming an increasingly important and relevant feature in the tourism industry, which entices customers and is being used in all process of customer booking.

17.15 ARTIFICIAL INTELLIGENCE (AI) IN HOSPITALITY

The hospitality industry comprises a broad categorization of various sectors within the service industry, including accommodation, food and drink services, theme parks, event planning, and travel and tourism [43]. It includes hotels, tourism companies and agencies, restaurants, and bars. It caters to various businesses and services associated with leisure and customer satisfaction [44]. Another feature of the hospitality industry is that it focuses on concepts of luxury, leisure, and pleasure, enjoyment, and experiences, and not concentrating on necessities and essentials. Hospitality goes hand-in-hand with tourism, contributing to a country's trade, business, infrastructural, and cultural development and revenue.

The hospitality sector uses AI to learn about customers using its data analytics so that the hotel personnel can better understand customer preferences. The technology can collect customer information about purchases, travel choices, journey patterns and itineraries, payment methods, hotel costing inquiries, and location references. This data is then utilized to provide a memorable experience for the guests while at the hotel.

17.15.1 DEVELOPMENT OF ARTIFICIAL INTELLIGENCE (AI) IN HOSPITALITY

The importance of AI is realized by hotel owners worldwide. The technology is paving its path steadily into the industry as the staff members remain under pressure to deliver the best service. Robots are welcoming guests at the hotel entrances and the reception. They are also used as concierge services. The first time a hotel used a robot was at the Hilton Hotel. "Connie," the robot there, helped provide information relating to a tourist destination to the customers who interacted with it [45]. This bot can interact and learn from humans. In fact, the more it speaks to humans, the better it performs.

Other areas like reservations, check-in, checkout, and monetary processes can all be handled by AI technology. Another example could be the Marriott Mobile App, through which the guests can directly access their rooms without going to the front desk. Allora, the world's platform to facilitate the direct booking, is AI-powered. Allora takes up bookings directly from guests to provide good experiences. She depends on learning models to assess vast quantities of data and find best configuration. Allora takes up individualized conversations with the guests to understand their requirements well with this information.

The American Gaming and Hotel Company, Caesars Entertainment Corporation, uses AI and Big Data to collect data about each guest according to their spending at each restaurant, spa, casino, and golf course. Accordingly, data is analyzed and tailored to serve the customer for a better customer experience. The organization offers Total Rewards Loyalty Program to retain customers, encourage them to spend more, and lead to customer loyalty and satisfaction.

17.16 USAGE OF ARTIFICIAL INTELLIGENCE (AI) IN HOSPITALITY

17.16.1 IMPROVING HOTEL OPERATIONS

Chatbots are increasingly being used in hotels to cater to online queries while the front desk executives handle hotel guests. This makes work handling easier and smoother, and both functions are taken care of simultaneously. Self-check-in kiosks are growing in popularity in the hospitality industry. These kiosks contribute to providing a hassle-free check-in and a good experience.

17.16.2 EFFECTIVE REVENUE MANAGEMENT

Artificial intelligence (AI) enables trend analysis to optimize hotel room rates. This is done by computing complex data and using the best rates across all channels to derive maximum revenue with less effort [37]. The cloud-based property management system (PMS) is another method of managing the bookings effectively. It is an automated and service-centric platform that helps in increasing work efficiency.

17.16.3 PERSONALIZATION REDEFINED

The hospitality industry is known for its personalized service feature. However, people are contented speaking with a chatbot on social media platforms while booking their hotels with the changing times. The more prominent entities even provide for integrated PMS in their operations to bring in service efficiency and a better position.

17.16.4 DATA ANALYSIS

The accomplishment of tasks like customer service, understanding guest preference, repeat customers, loyalty programs, and so on can be better achieved with the help of AI tools. Analysis of guest preferences to organize the room, accordingly, keeping in mind the budget of guests, and providing them with amenities accordingly to have them as regular clients can be handled by AI, making human labor less tedious.

17.16.5 MULTILINGUAL BOOKING EXPERIENCE

In some countries, many tourists visit hotels from different parts of the world. English is the most widely spoken language, though it is preferred that the front desk executive be abreast with more languages to cater to the requirements of guests speaking various languages. Chatbots can handle something which might always not be possible humanly. These machines can be tuned to speak a number of languages, helping the guests to be at ease and have the common queries answered easily.

17.16.6 ACCEPTANCE OF ARTIFICIAL INTELLIGENCE (AI) IN THIS INDUSTRY

Around 86% of hoteliers agree that AI can improve employee satisfaction. This is because the technology can handle tedious and monotonous tasks, letting the employees use their skills on high-value work, thus helping them to increase their knowledge and performance. However, it should always be borne in mind that these intelligent systems should be used only to better the old processes, not replace the human resource.

At the big hotels, AI is already contributing to curtail the employees' workload with its systems performing efficiently to take up the common queries of guests, booking rooms, handling food and beverage (F&B) orders, and so on [46]. Hotels that receive guests in large numbers can implement AI-enabled virtual assistants with natural language integration and self-learning capabilities to converse with the guests and help them with the standard requirements. Each interaction with the guests helps these systems to become better and improve the guest experience by engaging with them in a friendly manner.

It has been observed that technology usage is also helpful in speeding up work. The acceleration of call resolution times in many jobs has been noticed to improve. More than four minutes of calling time per interaction have been reduced while increasing accuracy up to 85% for the tasks on the queries received [46]. This has proved to have taken much pressure off from humans, enabling them to indulge in tasks requiring human attention.

At one of the hotel groups, the bots handle more than 2 million queries, taking up the workload of around 7,000 employees. The virtual concierge answers customer calls, reducing about 30% of calls at the human concierge desk. It is vital to accept the technology change that the hospitality industry is witnessing. Thus, the employees must understand the advantages of AI and how it can improve the guest and employee experience. They must understand that the technology allows them to focus on higher-level jobs by reducing the burden of the lower-level ones for them. One of the most talked-over topics about the effects of robotics and AI technologies on organizations is if the technology will succeed human workforce and, in the same relation, the changes in job responsibilities and the working relationships among employees and between employees and managers [47]. The adoption of this technology should help the employees and organization support receive professional training, including soft skills development, for better job prospects.

17.16.7 LIMITATIONS OF ARTIFICIAL INTELLIGENCE (AI) IN HOSPITALITY INDUSTRY

1. **Authentic Data:** A significant challenge in business to incorporate AI is generally not appropriately organized data. It is imperative to source the data efficiently for AI to utilize it for effective results.
2. **Operating AI Solutions:** Scientists and other staff, according to the study, are in short supply of who can determine the learning and training of hospitality professionals to handle the AI adoption skillfully.
3. **Cost:** Smart technology is expensive to incorporate as well as to maintain. It is for the individual hospitality entities to figure out if they can sustain themselves by adopting AI in different areas [41]. The computational costs for training data models, costs for repair, and regular maintenance are factors to be considered.
4. **Upgradation of Software Programs:** The ever-changing nature of business in hospitality would require constant updates and changes in the software programs that the AI programs would require. Critical data can also be lost in this manner. The process

of doing so is expensive and time-consuming. Thus, these points need to be accounted for while going for AI models.

Apart from the above, while implementing AI technology, it is essential to remember that customer privacy must not be breached. This would be against the ethics of hospitality.

Further, the adoption of AI could lead to an anticipated lack of transparency, which could ultimately threaten the business. The complex technology requires skilled hands to organize and utilize data properly to achieve maximum benefits from the system, rather than becoming a hurdle in operating the business.

17.16.8 SUMMARY – AI IN HOSPITALITY INDUSTRY

AI is a new technology. However, the trend is being noticed in the hospitality sector for some time now. The industry has even started capitalizing on the technology to reap benefits, monitory, and skill based. Whether it is the robots handling the front desk, offering concierge service, or answering customer queries, the technology is already driving the industry today. AI-enabled technology is the need of the hour, and businesses must opt for it in the strategic phase. Organizations realize the potential of the intelligent system to expand their businesses, bringing in changes for better service and increased customer satisfaction.

17.17 FUTURE OF ARTIFICIAL INTELLIGENCE (AI) IN THE SERVICE INDUSTRY

Today's customers understand that the global economy is going digital. They can connect whenever they want, get their appointments scheduled quickly, and have questions answered in a jiffy. The customers now generally converse with chatbots when they connect with a company for any product details. They get responses for general queries very quickly. AI and CRM are going hand-in-hand towards the future. The trend is gradually reaching out to all industries, with human interactions becoming virtual. Organizations realize the massive shift, adopting AI tools to collect information, take up customer issues and solve them with minimum errors.

17.18 CONCLUSION

AI has established a modern stage of intelligence to the customer service industry. As technology is advancing, more innovations are taking place. Organizations are laying more focus on providing better customer service by using AI algorithms. Many options are being explored to maintain a balance between technology and human interaction skills. The technology requires algorithms that AI engineers prepare to provide intelligent customer support. This is acquired from the data collected by customers' behavior on various online media, such as social media, apps, websites, and a lot more. The usage of AI needs to be in the right direction to benefit humanity and not threaten human resources for work. With the role of AI and machine learning in the upcoming times, it is the right time to furnish human resource skills and embrace this technology.

KEYWORDS

- **artificial intelligence**
- **augmented reality**
- **data analysis**
- **education**
- **healthcare**
- **hospitality industry**
- **online reputation management**
- **service industry**
- **tourism**
- **virtual reality**

REFERENCES

1. Russell, S., & Norvig, P., (2016). *Artificial Intelligence*. California: CreateSpace Independent Publishing Platform.
2. Frankenfield, J., (2021). *Investopedia*. Retrieved from www.investopedia.com (accessed on 07 September 2023).

3. Kumar, P., Jain, K., & Kumar, D., (2021). *Artificial Intelligence and Global Society.* Bhopal: CRC Press.
4. Lansity, M., & Lakhani, R., (2020). *Competing in the Age of AI.* Harvard: Harvard Business Review Press.
5. Kaplan, J., (2016). *Artificial Intelligence.* California: Oxford University Press.
6. Rouhiainen, L., (2020). *Artificial Intelligence.* Valencian Community: Createspace Independent Pub.
7. OECD, (2019). *Artificial Intelligence in Society.* Paris: OECD Publishing.
8. Pupillo, L., Fantin, S., Ferreira, A., & Polito, C., (2021). *Artificial Intelligence and Cybersecurity.* CEPS.
9. Carbo, S., Fernández, R. F., & Solas, C. P., (2016). *Bank Funding, Financial Instruments and Decision-Making in the Banking Industry.* Valencia: Springer International Publishing.
10. Digalaki, E., (2019). *The Impact of Artificial Intelligence in the Banking Sector & How AI is Being Used in 2020.* Business Insider.
11. Suhel, S. F., Shukla, V. K., Vyas, S., & Mishra, V. P., (2020). Conversation to automation in banking through chatbot using artificial machine intelligence language. In: *2020 8th International Conference on Reliability, Infocom Technologies and Optimization (Trends and Future Directions) (ICRITO) 2020* (pp. 611–618). doi: 10.1109/ICRITO48877.2020.9197825.
12. Boobier, T., (2020). *AI and the Future of Banking.* England: Wiley.
13. Bansal, A., (2021). *How AI And Machine Learning Are Transforming the Banking Industry.* Business World.
14. Sastry, N. L., (2020). *Artificial Intelligence in Financial Services and Banking Industry.* Mumbai: Idea Publishing.
15. Thekkethil, M. S., Shukla, V. K., Beena, F., & Chopra, A., (2021). Robotic process automation in banking and finance sector for loan processing and fraud detection. In: 2021 *9th International Conference on Reliability, Infocom Technologies and Optimization (Trends and Future Directions) (ICRITO), 2021* (pp. 1–6). doi: 10.1109/ICRITO51393.2021.9596076.
16. Chishti, S., (2020). *The AI Book.* England: Wiley.
17. Coulbeck, S., (2012). *The Multinational Banking Industry.* New York: Taylor & Francis.
18. Marketsandmarkets, (2018). AI in Education Market by Technology (Deep Learning and ML, NLP), Application (Virtual Facilitators and Learning Environments, ITS, CDS, Fraud and Risk Management), Component (Solutions, Services), Deployment, End-User, and Region – Global Forecast to 202. Marketsandmarkets.
19. Knox, J., Wang, Y., & Gallagher, M., (2019). *Artificial Intelligence and Inclusive Education.* Edinburgh: Springer Singapore.
20. Java, S., Mohammed, H., Bhardwaj, A. B., & Shukla, V. K., (2021). Education 4.0 and web 3.0 applications in enhancing learning management system: Post-lockdown analysis in COVID-19 pandemic. In: Sandeep, K., Deepmala, S., Zdzislaw, P., Alka, M., & Mary, J., (eds.), *Knowledge Management and Web 3.0: Next Generation Business Models* (pp. 85–110). Berlin, Boston: De Gruyter. https://doi.org/10.1515/9783110722789-006.

21. Analytics Insight, (2021). *Artificial Intelligence in 2021 – The Developments so Far.* Retrieved from: www.analyticsinsight.net (accessed on 07 September 2023).
22. Holmes, W., Bialik, M., & Fadel, C., (2019). *Artificial Intelligence in Education.* London: Independently Published.
23. Salajan, F. D., (2019). *The Educational Intelligent Economy.* Mexico City: Emerald Publishing Limited.
24. Cameron, M. R., (2019). *A.I. – 101: A Primer on Using Artificial Intelligence in Education.* California: Exceedly Press.
25. Muldner, K., Bittencourt, I., Lickin, R., Millán, E., & Cuckurova, M., (2020). *Artificial Intelligence in Education.* New York: Springer International Publishing.
26. Habib, M., (2019). *Revolutionizing Education in the Age of AI and Machine Learning.* Cairo: IGI Global.
27. Aoun, J., (2017). *Robot-Proof.* Massachusetts: MIT Press.
28. Davenport, T., (2019). The potential for artificial intelligence in healthcare. *Future Healthcare Journal, 6*(2), 94–98.
29. Athota, L., Shukla, V. K., Pandey, N., & Rana, A., (2020). Chatbot for healthcare system using artificial intelligence. In: *2020 8th International Conference on Reliability, Infocom Technologies and Optimization (Trends and Future Directions) (ICRITO), 2020* (pp. 619–622). doi: 10.1109/ICRITO48877.2020.9197833.
30. Fakoor, R., Ladhak, F., Nazi, A., & Huber, M., (2013). Using deep learning to enhance cancer diagnosis and classification. *The 30th International Conference on Machine Learning*, (p. 7). Arlington.
31. Richins, S. M., (2015). *Emerging Technologies in Healthcare.* New York: Taylor & Francis.
32. Nordlinger, B., Villani, C., & Rus, D., (2020). *Healthcare and Artificial Intelligence.* Paris: Springer International Publishing.
33. Bohr, A., & Memarzadeh, K., (2020). *Artificial Intelligence in Healthcare.* Colorado: Elsevier Science.
34. Chang, A. C., (2020). *Intelligence-Based Medicine.* California: Elsevier Science.
35. Kabene, S. M., (2010). *Healthcare and the Effect of Technology.* Paris: IGI Global.
36. Jacques, B. G., (2020). *Handbook of E-Tourism.* Las Palmas: Springer.
37. Webster, C., & Ivanov, S., (2020). *Robots in Travel, Tourism, and Hospitality: Key Findings from a Global Study.* Sofia: Zangador.
38. Monfared, M., Shukla, V. K., Dutta, S., & Chaubey, A., (2022). Reshaping education through augmented reality and virtual reality. In: Tavares, J. M. R. S., Dutta, P., Dutta, S., & Samanta, D., (eds.), *Cyber Intelligence and Information Retrieval: Lecture Notes in Networks and Systems* (Vol. 291). Springer, Singapore. https://doi.org/10.1007/978-981-16-4284-5_55.
39. Ansari, S. Z. A., Shukla, V. K., Saxena, K., & Filomeno, B., (2022). Implementing virtual reality in the entertainment industry. In: Tavares, J. M. R. S., Dutta, P., Dutta, S., & Samanta, D., (eds.), *Cyber Intelligence and Information Retrieval. Lecture Notes in Networks and Systems* (Vol. 291). Springer, Singapore. https://doi.org/10.1007/978-981-16-4284-5_49.
40. Schegg, R., & Inversini, A., (2016). *Information and Communication Technologies in Tourism 2016.* Vaud: Springer International Publishing.

41. Webster, C., & Ivanov, S., (2019). *Robots, Artificial Intelligence and Service Automation in Travel, Tourism and Hospitality.* Muncie: Emerald Publishing Limited.
42. Gajdošík, T., & Marciš, M., (2019). *Artificial Intelligence Tools for Smart Tourism Development.* Springer.
43. Verma, A., Vinod, K. S., & Robin, S., (2021). Convergence of IOT in tourism industry: A pragmatic analysis. *Journal of Physics: Conference Series, 1714*(1). IOP Publishing, 2021.
44. Brotherton, B., (2012). *International Hospitality Industry.* Oxfordshire: Taylor & Francis.
45. Armbruster, J., (2019). *Artificial Intelligence in the Hospitality Industry.* North Rhine-Westphalia: GRIN Verlag.
46. Zabin, J., (2019). *Artificial Intelligence: Working Hand in Hand with Hotel Staff.* Hotel Technology News.
47. Qiu, H., Li, M., Shu, B., & Bai, B., (2019). Enhancing hospitality experience with service robots: The mediating role of rapport building. *Journal of Hospitality Marketing & Management, 29*(3), 247–268.

CHAPTER 18

Technological Implementation in the Service Sector: A Case Study

E. EZE IFEANYICHUKWU

Diversity and Inclusion Specialist, Royal Springs International Abuja, Nigeria

ABSTRACT

The service industry is a crucial industry recognized as the biggest of the three sectors of the economy. Implementing technology in this industry has been the driving force for industrial advancement. In contrast, this industry's disproportionate adoption of technology explains the disparity between developing and developed worlds. Artificial intelligence (AI) has been recognized as the newest technology disrupting and driving innovation in the service industry, with the quest for superior business experience and customer service driving its adoption. However, AI is not the only technology driving transformations in the service industry but also a constituent of innovative technologies, such as search, chatbots, cloud computing, augmented reality (AR) and virtual reality (VR), mobile, social, and the IoT. In addition, this chapter advocates for the fourth component of cloud computing to be named diversity as a service (DaaS). Technology is the most significant enabler of globalization. Globalization is the most fantastic driver of diversities – creating opportunities, virtual, and physical, for people from different cultural, religious, racial, geographical, and other backgrounds to interact. However, this confluence of diversities is leading to unsavory human experiences giving rise to racial protests, gender inequalities, underrepresentation of ethnic minorities, and others.

However, implementing technology in diversity management could be the elixir to these problems.

The chapter seeks to explore the implications of technology implementation in the service industry to ascertain whether the economic impact of technology in the service industry influences its adoption and how to encourage technology adoption in the service industry. The chapter also analyzes whether technology implementation in the service sector is economical to provide organizations with a decision-making tool for further exploring the industry and making experimental projections of technology's prognosis in the service industry to encourage research and scientific activity. This is because many or most of the literature on AI are projections, simulations, and assumptions emphasizing how the implementation of AI will increase unemployment and change the labor landscape in ways that may not be familiar to many. We adapted the core generalizations from some of these reports to evaluate these projections. Academic and research reports from the sciences, engineering, and medicine on information technology (IT) and the U.S. workforce into an analytical framework because it embodies the core assumptions of most of the literature on the subject. We then used the analytical framework to analyze secondary data from those who have implemented these technologies to see whether the implementation of technology in the service sector was economical or not and whether economic factors affected the adoption of the technology. Finally, we explored possible extensions of existing knowledge as experimental projections to encourage further research.

Furthermore, the chapter explored three critical sectors of the service industry as case studies, including the Healthcare, Logistics, and Entertainment sectors, and attempted an economic analysis of the implementation of technology in these sectors and projected experimental investigation of the consequences of implementing technology in these sectors in the future. It was noticed that the implementation of technology in the service industry is gaining increased adoption with signs of growing adoption into the future. Economically, the performance of technology increases productivity, lowers costs, and transforms the way work is done and changes the labor landscape, with machines gaining greater involvement and acceptance as co-workers with human counterparts. This shows the economic benefits of implementing technology in the service sector in increasing its adoption. Most importantly, it was established that different levels of education, not technology, are the causes of income inequalities.

Generally, the implementation of technology increases the efficiency and effectiveness of service delivery in these sectors, leading to labor efficiency, increased productivity, and cost-effectiveness.

Experimentally, the chapter projects that we are entering the era of flying cargoes and rocket-like cargo ships. The chapter concludes with claims that though machines will see greater adoption in the entertainment industry, they will not substitute human creativity in serving entertainment, and humans will always cherish entertainment served by humans and treat machine-generated entertainment as fiction.

18.1 INTRODUCTION

The service industry has been touted as the biggest of the three sectors of the economy. It is analogous to the last runner in a relay race that defines and seals the fate of all the successive runners. The newest advance in this sector revolves around AI and how its implementation has shown to have the potential to revolutionize not just the service industry but all the sectors of the economy.

In this chapter, we shall explore the implications of the implementation of technology in the service industry to determine whether the economic implications of technology in the service sector affect the adoption rate of technology in the service sector and how to stimulate increased adoption. Most of the reports and literature based on Artificial Intelligence (AI), indicate that AI may bring benefits in the future, but this can also be one reason for unemployment.

Historically, industry 1.0 and 2.0 were the eras of the introduction of technology, resulting in increased production. In contrast, industry 3.0 was the era of technological production and customer service improvement, with industry 4.0 now the era of advancement of technological productions and customer service, with AI as the mainstay. It is, therefore, pertinent to note that the implementation of technology has always been driven by the quest to achieve better customer satisfaction and enhanced business operations. As noted by [1], "the provision of superior customer service experience was the Highest driver of AI adoption. Moreover, authors [1] further claims there is a direct link between superior business outcome and adoption of AI."

As a result, artificial intelligence as a service (AIaaS) is witnessing rising investments for its adoption and workflow optimization techniques. Companies are exploring new ways of increasing their profit margins with lower operational costs. IaaS is segmented by organization's size, type of technology, service provided, type of cloud used, vertical, regional characteristics, competitiveness, forecast, and opportunities. Therefore, AI is the newest bride in technological innovation, disrupting the global economy and people's way of life. The Artificial global intelligence (AI) market is projected to grow from USD 47.47 billion in 2021 to USD 360.36 billion in 2028 at a CAGR of 33.6% in the 2021–2028 period [2]. Like in other industrial eras, it has economic implications in that it is changing the organization and function of labor, leading to cost-effectiveness, and increased productivity.

Even though AI enjoys pride in the technological innovation revolutionizing the service industry, it is not the only technology influencing or driving the transformations in this industry. AR, virtual reality (VR), search, and cloud computing, chatbots, mobile, social, and IoT are primary ways technology is changing the service industry [3].

18.2 IMPACT OF SEARCH ENGINES ON THE SERVICE INDUSTRY

Search engines have become the first port of call when customers want to explore the range of service offerings available to them. Search technology has drastically changed the way companies, organizations, and businesses relate with their customers and each other. Because of the strategic nature of the search functionality, businesses, and even governments are now employing SEO experts to make sure their communications appear on the top ranks of the search engine results page (SERP). As a result, not searchable communications have been deemed a waste of time and resources and a huge loss to the organization. Bill [4] asserts "that Google's success cannot be attributed to chance or luck, but to the company's ability to radically innovate their technologies and services, business model, organizational culture and a dynamic leadership style which has set it above others in the search services ecosystem." Similarly, [5] highlights four essential reasons why site search is crucial for online retailers as follows:

- A user-friendly website search functionality, which helps in easy navigation, has become a critical component of any organization's website and digital strategy. This website search function is also an invaluable business asset, leading companies to increase their investments in search, given the overwhelming business results the search functionality delivers businesses.

- Companies get critical and strategic business data and customer behavior insights from customers' search terms on their websites. These data help companies learn more about their customers and customize their offerings.
- Because of enormous benefits from search, companies are now investing more in this technology to stay competitive.
- Strategic deployment of search data could lead to increased conversions and customer retention rates.

The diagram in Figure 18.1 shows that all aspects of search technology, be it direct, organic, etc., is critical to business and service visibility.

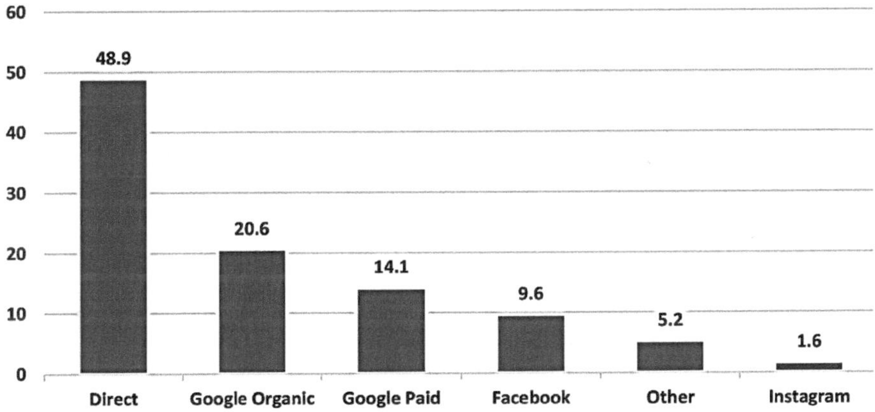

FIGURE 18.1 Share of commerce traffic generation in percentage [6].

18.3 IMPACT OF CHATBOTS ON THE SERVICE INDUSTRY

According to Ref. [7], a chatbot is a self-operating system that can imitate person-to-person dialog via voice or text messages. Chatbots are now largely deployed as cost-saving and efficient alternatives to one-on-one interactions [8]. Siri and Alexa are beautiful examples of chatbots. Chatbots are increasingly dominating the customer service landscape because of their cost-effectiveness and round-the-clock availability.

Chatbots have revolutionized customer service by providing basic and vital information to clients 24/7 and also present opportunities for companies to see gaps in their customer services.

Authors [9] further confirms the role of chatbots in the service industry by stating "that market size will continue to grow as chatbots continue to lower operating costs for businesses. They further claim that even though chatbots could be deployed in other business sectors as payments, services, and processing units, that revenue from services will always command a greater share of the market share." Chatbots render customer services at a reduced cost to organizations while increasing the revenue organizations generate from them. In addition, Tables 18.1 and 18.2 further detail the impact of chatbots in the service industry.

TABLE 18.1 Advantages of Chatbots to Customers

Benefits	Percentage (%)
24/7 service	64
Undelayed response	55
Receiving feedback to questions	55
Lack of difficulties in communication	51
Rapport and accessibility	32

TABLE 18.2 Consumer's Usage of Chatbots

How Consumers Are Using Chatbots	Percentage (%)
Receiving instant response during emergencies	37
Settling complaints or problems	35
Receiving detailed feedback	35
Getting access to a human customer service agent	34
Booking reservations	33
Payment of bills	29
Buying basic commodities	27
Obtaining insights for purchases	22
Others	14

From Table 18.3, it is clear that chatbots are becoming increasingly indispensable in providing customer services in the service industry.

TABLE 18.3 Artificial Intelligence Chatbots Acceptance by Customer (by Industry)

Industry	Percentage (%)
Online Marketing	34
Health services	27
Telecoms	25
Banking services	20
Financial guide	20
Insurance services	15
Government	10

Source: Ref. [10].

18.4 IMPACT OF CLOUD COMPUTING ON THE SERVICE INDUSTRY

Cloud computing provides on-demand computing resources such as databases, storage, analytics intelligence, software, and networking to users when they need it, without users having to own or manage the infrastructure. In other words, cloud computing is the heart or engine of AI and the latest innovation in data storage that makes it more efficient and effective to manage data. These services are software as a service (SaaS), infrastructure as a service (IaaS), and platform as a service (PaaS). However, in this chapter, we want to advocate for the fourth component of cloud computing to be DaaS; details of DaaS will be given at the end of this chapter.

SaaS makes data easily accessible and shareable online. Examples of this service include Salesforce, Dropbox, HubSpot, and others. IaaS is a digital infrastructure that enables instant computing over the internet. Instances of IaaS are Google Compute Engine, Microsoft Azure, Cisco MetaCloud, among others. Then, PaaS is a holistic environment that makes it efficient to develop and deploy cloud-based resources from simple apps to complex applications. Examples of PaaS include AWS Beanstalk, OpenShift, Force.com, etc.

Cloud computing has some drawbacks, as it were, but the advantages far outweigh the flipside. Therefore, some of the advantages include but are not limited to flexibility, rich customer experience, lower costs, scalability, and better security, among others.

18.5 IMPACT OF AUGMENTED AND VIRTUAL REALITY (VR) ON THE SERVICE INDUSTRY

VR is the use of technology to produce believable real-life experiences virtually, whereas AR is the use of computer-generated content to augment real-life experiences.

In the health sector, the impact of AR is manifest in augmented diagnosis and augmented surgery, which has made diagnosis and surgery a lot easier and more efficient. Other sectors that AR is revolutionizing include entertainment, military, manufacturing, supply chain, education, etc. Figure 18.2 shows the adoption rate of AR/VR across industries.

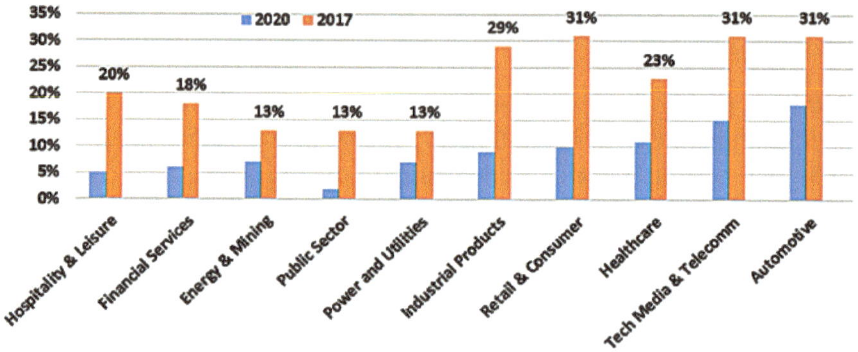

FIGURE 18.2 Percentage of companies investing in AR by industry.

Source: Adapted from Ref. [11].

18.6 IMPACT OF MOBILE TECHNOLOGIES ON THE SERVICE INDUSTRY

The impact of mobile on the service industry is pervasive and enormous. It has redefined business processes in many ways. Businesses now communicate with their clients anytime through instant messaging apps. Because many people use mobile services, businesses ensure that their websites are responsive and adaptive. The use of customized mobile apps is now the easiest way to visit a business, access or use their services from any part of the globe. All these important apps are hosted on mobile devices.

Geolocation services, too, now allow companies to know customers who are close to their location and thereby serve them customized services. According to Ref. [12], "m-commerce volume would grow to 25.5% compound annual growth rate from 2019 to hit $488.0 billion, or 44% of e-commerce, in 2024. The driving force of this growth, he claims, are smartphones and tablets."

18.7 IMPACT OF SOCIAL ON SERVICE INDUSTRY

Social media's ability to gather billions of people on a single platform like Facebook and Twitter has made it the single most populous market in the world. Furthermore, this makes it possible for ads to go viral using the share button. More so, its ability to gather people in groups around a common interest makes it possible for businesses to have access to groups or many people interested in their product offerings. Social media also helps these businesses to know more about the discussions on their products, which helps them refine or improve their offerings to suit popular demand. It also provides avenues for mutual and robust interactions between organizations and their customers. Many customers today prefer doing business with companies with a solid social media presence. Many customers prefer customer services served on social media, as the chart in Figure 18.3 shows.

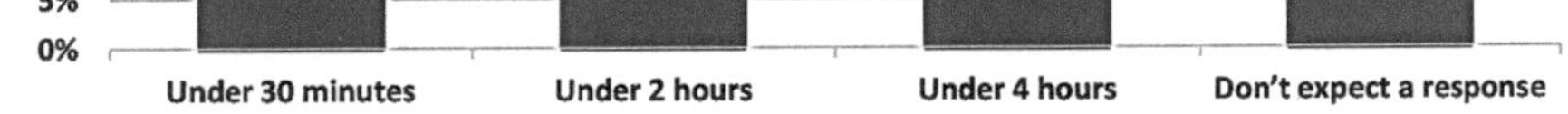

FIGURE 18.3 Customers waiting preferences from their brands.

Source: Adapted from Ref. [6].

From the responses above, it is clear that more customers prefer to receive responses on social media within 30 minutes of making a request. This explains why many organizations are now employing social media marketers to cater to the teeming customers who follow them on social media.

18.8 IMPACT OF IOT ON THE SERVICE INDUSTRY

Internet of Things (IoT) refers to the connectivity between devices and digital technologies made possible by the internet and sensors. These connected devices can transmit or share data without human assistance. Every IoT component possesses a unique identifier (UID) which makes it identifiable. For now, IoT applications are five-fold as follows:

1. **Consumer IoT:** These include such applications as light fixtures, office, and home appliances, and voice assistance for the elderly.
2. **Commercial IoT:** This includes the application of IoT in healthcare and logistics sectors, involving the use of products like smart pacemakers, monitoring devices, and inter-vehicular communications.
3. **Industrial Internet of Things (IIoT):** These are systems that are digitally controlled, with capabilities for analyzing statistical data, for use in agriculture and industries.
4. **Infrastructure IoT:** These are devices that enable the linking of smart cities via the deployment of sensors, apps, and management systems.
5. **Military Things (IoMT):** This involves the use of digital innovations in military activities. For example, the use of humanoids for reconnaissance and surveillance operations and the use of wearable biometrics for combats.

People now spend a significant chunk of their time online, from watching online TV like Netflix to browsing social media sites like Facebook, Instagram, WhatsApp, etc., to using blogs and websites like Amazon.com, google.com to streaming videos on YouTube and digital devices. We leave digital footprints that can reveal our preference for different services, our choice destinations, food, fashion delights, and many more through these activities. With these humongous amounts of data generated and insights made, IoT enables service providers with better insights on how best to render their services, automate their workflows, and increase customer satisfaction. According to Ref. [13], "the IoT revolution is driving innovations in the form of smart homes, offices, and factories and many other domains. Emphasizing that modern-day offices are dependent on connected devices such as biometric scanners, air conditioners, and others – to create a more supportive and healthier space for workers, as well as cutting down on energy usage and demands." Today, many businesses rely on data from cameras in the service halls and environment to interpret customer behavior satisfaction towards their services (Figure 18.4).

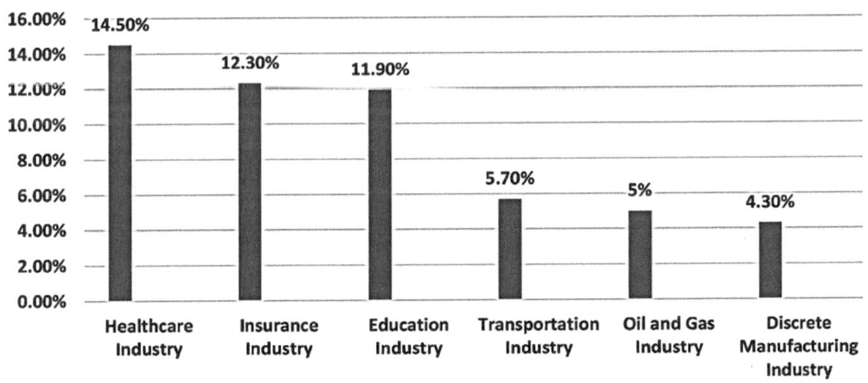

FIGURE 18.4 Growth rate of IoT spending in 2020 [14].

18.9 ANALYTICAL FRAMEWORK

In choosing an analytical framework, we want a framework that can address the interactions between the economy and technology, especially in the service industry, with global applicability. This framework is authored by a group of experts with a global reputation, which should not be more than four years old. As a result, we settled for the report by technology experts, scientists, and medical experts in the United States of America because, according to the authors, the reports analyze the relationships, interactions, and impacts of technology on society and the resultant effects in terms of economic, technological, and societal trends going forward. We shall, for analytical reasons, focus only on the generalized core assumptions in the report that summarizes the implications of the implementation of technology and adapt it for our use.

Given the reasons mentioned earlier, we chose the report of Ref. [15] because of the comprehensiveness of their study on how ICT impacts society and work economically, the global composition of the members, and its potential for global application and correlation with our case studies. These core generalizations are namely:

1. The key drivers of advancements in productivity are factors such as diffusion, innovation, and adoption of technology, and that the most relevant technology of this era is an IT. However, the existence of technology is not adequate for productivity growth, claiming that productive use of technology requires complementary skills among the workforce, and alignment of institutional and business processes, and organization of work, which, according to them, is very expensive time-consuming.
2. That technology has revolutionized employment by initiating machine-driven task-performance capabilities, which can continue to do so. They assume this would lead to job losses, obsolescence of existing jobs, and the need for entirely new ones, including occupational shifts and the need for new skill sets. In contrast, the percentage of working people has reduced in the past 20 years; shifts in and across employment will be more economically viable than the variations in the total proportion of employment.
3. Those latest computerized technologies seem to have added to the rising income inequality. They may likely continue, provided they keep replacing skills and tasks hitherto identified with low-wage, menial, or middle-wage occupations. The jobs are left to demand more intellectual and analytical skills, or they render personal services with little or no economic importance at the moment. These variations will most likely be manifest in wages and incomes, even though they might be noticeable to some degree in the number of hours worked and employment generally.
4. That cost efficiency is relative to other factors.

18.10 CASE STUDIES

In the following case studies, we will be considering how the implementation of technology is revolutionizing some critical sectors of the service industry. Because of scope, we will limit ourselves to three important sectors: Logistics, Healthcare, and Entertainment.

18.10.1 LOGISTICS SECTOR

The logistics sector is witnessing substantial technological innovations powered by AI. From inventory, packing, and sorting to delivery, the entire logistics process is witnessing technological transformations. Today, smart warehousing is becoming the order of the day, and many businesses are adopting the new model. If any sector of the economy needs radical transformation, it is the logistics industry. Global business is already happening at a speed nearing the speed of light. E-commerce has transformed global business magically, thanks to the implementation of technology. Consumers from anywhere in the world could shop anywhere in the world from their mobile phones, tablets or laptops.

Therefore, the implementation of AI in logistics has shown to be a Daniel come to justice. AI is revolutionizing almost every aspect of logistics, from warehousing to transportation and delivery (Figure 18.5). Today, we have automated warehouses, automated order fulfilments and delivery. Customer satisfaction is now peeking. However, despite these developments, the adoption of these technologies is still in the lower percentile, as we shall see in the graphs (Figure 18.6).

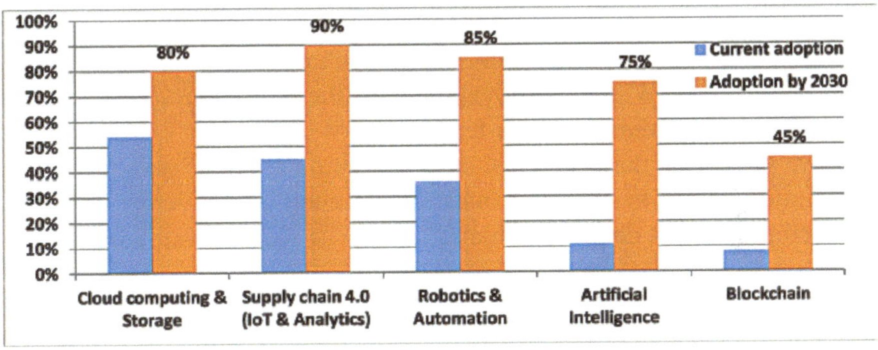

FIGURE 18.5 Adoption of AI in warehouse automation [16].

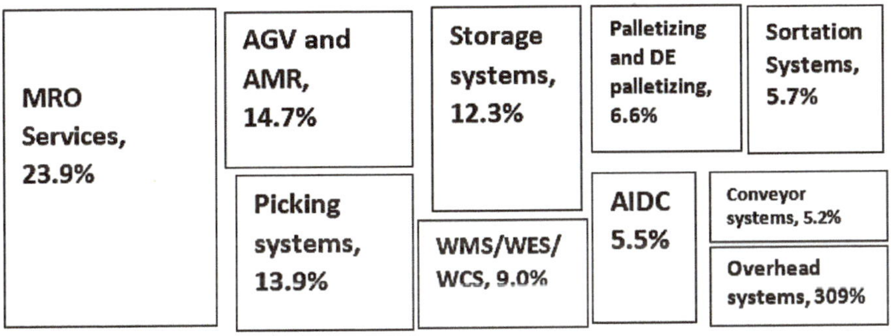

FIGURE 18.6 Projected global warehouse automation market breakdown by technology by 2025 [16].

18.10.1.1 CAINIAO SMART LOGISTICS

Cainiao Smart Logistics is a modern logistics company that leverages big data and Cloud computing technologies to drive a platform. This integrated system can efficiently link third-party logistics firms and e-commerce businesses for enhanced service and efficient delivery. The company recently launched Xiao g, a Smart delivery cart. Xiao G is one meter tall and can load up to 10 to 20 parcels at a time. It has the capacity for intelligent route analysis that helps it receive parcels from nearby distribution centers, triggers alarms when someone tries to take a parcel that is not theirs and can use elevators. Cainiao goal in launching xiao g was to achieve 24-hour delivery of goods within China, and 72 hours globally [17, 18].

18.10.1.2 JD.COM SMART LOGISTICS

JD Logistics is the first company to launch a 5G-powered Smart logistics park in China and build automated warehouses with last-mile drone delivery services. With these developments, the traditional logistics landscape is being revolutionized by technology [4, 19].

"We are taking the lead in robotics and automation, as we are the first company globally to deploy drones for commercial deliveries and have evolved unmanned delivery vehicles, stores, as well as warehouses. And since we have a platform that generates up to 31 petabytes daily, we are developing sophisticated AI and machine learning capabilities that will provide unparalleled customer behavior analytics in China," said [20].

18.10.1.3 AMAZON ROBOTICS (KIVA ROBOTICS)

Amazon robotics (otherwise known as kiva robots) have revolutionized how the e-commerce giant fulfills customers' orders. Amazon started using robots in 2012 and have achieved large scale order fulfilment. Amazon robots can carry shelves of products from one workstation to another and read barcodes designed to guide its movement. Amazon uses its robots to achieve faster order fulfillment, automate its work processes, and lower costs. Amazon workers have eliminated 12 miles of walking on concrete floors daily by this technological innovation [21].

18.10.1.4 ECONOMIC ANALYSIS OF TECHNOLOGY IMPLEMENTATION IN LOGISTICS

1. That the critical drivers of advancements in productivity are factors such as diffusion, innovation, and adoption of technology; and that the most relevant technology of this era is an IT. This factor of innovation, diffusion, and adoption is currently manifest in Cainiao open-source platform that facilitates collaboration among 3,000 logistics companies, 3 million couriers, including local and international partners. The Cainio system has been likened to the Android system. It is an open-source system that allows different logistics companies from around the globe to key in, enabling faster sharing of real-time data and intelligent route analysis, which increases cost-and-time efficiency delivery. Now, this system delivers parcels that generally take three days to deliver in just 24 hours at a reduced cost. However, this system is not a radical transformation but an extension of existing technology to other areas where transformations are needed to increase productivity. So, the Cainio system is relatively expensive to implement since it makes it easier for existing logistics systems to leverage it and increase productivity. Another instance of technology achieving increased efficiency and productivity in the logistics sector is the case of increased efficiency in the operations of Shentong (a Chinese company). The company uses self-charging robots that operate 24/7 with a capacity to sort a maximum of 2,00,000 parcels a day, thereby saving about 70% of the costs a human-based sorting would have required [17, 22].
2. That cost-efficiency is relative to other factors. The Xiao g delivery cart has contributed to the reduced logistics cost, especially the last mile delivery costs. It is achieving this through the reduction of delivery and packaging costs. For instance, "a DHL parcel that weighs less than 0.5 kg could be sent to London from Shanghai at the cost of about $100 presently and normally takes five days. However, xiao g is to send a typical 1 kg package anywhere in China within 24 hours for around 30 cents, and would deliver it for $3 any location in the world in not more than 72 hours." This system is also environmentally friendly and saves enormous amounts of resources spent on packaging. "Savings from

e-shipping labels alone amount to more than 400 billion pieces of paper and cancels carbon emissions per year worth a billion kilograms" [22].

3. The economic impact of technology on logistics is seemingly more visible in Amazon. At Amazon, robots have been said to complement humans, thereby reducing their workload. Amazon has created a work environment where robots and humans work together to fulfill orders in record time. According to Amazon, "Amazon presently operates 175 fulfillment centers globally. People work together with robots to pick, transport, stow and sort packages. Automation is making it easier for workers to perform their more laborious and tedious jobs by taking over such jobs such as heaving loads of inventory and moving pallets around." By implication, the new labor landscape is such that machines take over strenuous jobs that lead to inefficiencies. Humans have machines as co-workers, with the added requirement that humans who work in such warehouses have the requisite skills to work with robots. These skills are not rocket science but similar to the general digital knowledge needed to use a smartphone or just a slight extension. Craig [23] notes that "increase in the quality of labor comes from more training and qualitative education of the employees. Investments in computers, robotics, machines, and other items influence growth in productivity, and better outputs are driven by capital investments. Then, total factory productivity is achieved from the synergies of efficient interactions between labor and capital." An increase in productivity due to an increase in labor efficiency also leads to a gross domestic product (GDP) increase. Therefore, the technological innovations in the logistics sector are enabling healthier employment opportunities. In other words, technology is replacing strenuous, inefficient jobs with more efficient and more innovative options. In 2014, for instance, the introduction of the Cainio system in the logistics sector enabled the employment of 1.5 million personnel located in 600 cities by Cainio's partners [24].
4. Those latest computerized technologies seem to have added to the rising income inequality and may likely continue. This point is not readily visible in companies that have recently implemented these technologies, like Amazon, Cainiao, and JD.com. These

technologies are simply equipment for working efficiently. The disparity in education is instead the main decider of income inequalities as [23] above states: "increase in the quality of labor comes from more training and qualitative education of the employees. Investments in computers, robotics, machines, and other items influence productivity growth, and better outputs are driven by capital investments. Then, total factory productivity is achieved from the synergies of efficient interactions between labor and capital."

18.10.1.5 EXPERIMENTAL ANALYSIS OF AI IMPLEMENTATION IN LOGISTICS

This section clarifies that we are not making far-flung projections about technological developments in the logistics sector, but rather a near-term prognosis of technology in the sector based on existing knowledge. With many companies like J. D and others already deploying unmanned vehicles and drone technologies in delivery services, coupled with the advances NASA and Space X are making in space exploration, it is only expedient to encourage the extension of knowledge from these areas to solve transportation issues in the logistics sector. Today, ships remain the largest carriers of cargo in the world. It takes an average of 21 days to transport cargo from the United Kingdom, for instance, to South Africa. This delay in the delivery of goods from vendors to buyers is the most significant problem in the logistics sector.

In order to solve this problem, scientists should be working on the possibilities of using AI to adapt the speed of light in transport vehicles. Once this is done, in addition to the existence of infrastructures for drone transportation, we will enter the era of rocket ships. These unmanned delivery vehicles can deliver goods to any part of the globe within 24 hours.

18.10.2 HEALTHCARE SECTOR

Technology has been revamping and revolutionizing how services are rendered in the healthcare sector for centuries now with heavy economic implications. Technological innovations endear themselves to their users because they guarantee efficiency, cost-effectiveness, and the potential for customizable service applications. "Misdiagnosis and error in medical reviews accounted for about 10% of all deaths in the U.S. in 2015. As a result, AI's potential for increasing diagnostic accuracy is one of its most exciting applications in the health sector," Says [25]. Some examples of AI innovations that are revolutionizing service delivery in healthcare are Path AI (more accurate cancer diagnosis with AI), Enlitic (AI and deep learning for actionable insights), Freenome (Earlier cancer detection with AI), Bioxcel Therapeutics (AI in Biopharmaceutical development), and Berg Health (Treating rare diseases with AI) are very popular [26, 27].

Breakthrough AI technologies are many and varied in application, and their numbers are still growing. For now, existing A1 innovative technologies in the healthcare sector can be grouped under the following categories:

- Robot-assisted surgeries;
- Virtual nursing assistants;
- Administrative workflow assistance;
- Fraud detection;
- Dosage error reduction;
- Connected machines;
- Clinical trial participant identifier;
- Preliminary diagnosis;
- Automated image diagnosis;
- Telemedicine;
- Internet of medical things;
- AR/VR/MR in healthcare [28, 27].

Although AI technological innovations have many breakthroughs in the healthcare sector, it also has issues regarding its general acceptability, policies, and regulations for implementation, data privacy and safety issues [26]. However, Blockchain technology promises to be the answer to the issues above. "In addition to regulations like HIPPAC, coherence, safety, availability, and flexibility of data are characteristics that are highly sought after in the face of the COVID-19 global pandemic in 2020. For better communication between healthcare professionals and patients, Blockchain technology can be considered, achieve productive results" opines [29].

AI in the health market is a potentially huge market, with room for rapid growth, given the pervasive nature of the technology. The market is projected to be worth about $105 billion by 2026 [30]. The author's [30] analysis of 10 AI applications according to adoption possibilities and worth to the health economy is shown in Table 18.4.

TABLE 18.4 Approximately Health Economy

Application	Value (billion)
Robot-assisted surgery	$40 b
Virtual nursing assistants	$20 b
Administrative workflow assistance	$18 b
Fraud detection	$17 b
Dosage error reduction	$16 b
Connected machines	$14 b
Clinical trial participant identifier	$13 b
Preliminary diagnosis	$5 b
Automated image diagnosis	$3 b
Cybersecurity	$2 b
Total	= ~ **$148 b**

The next Table 18.5 shows the social economic impact of AI in healthcare.

TABLE 18.5 Social Economical Impact: AI in Healthcare

Resources Saved	Duration	Amount	Application	Amount Saved
		In Thousands	**Wearables**	313,000
Lives (per person)	Annually	380,000.00	Monitoring	42,000
		403,000	Imaging	41,000
Cost (€)	Annually	**In Billions**	**Wearables**	€50.6 billion
		170.0 to	Monitoring	€47.7 billion
		212.4	Real-world data	€38 billion
Time (hrs.)	Annually	**In Millions**	**Virtual health**	1,154 million hrs.
		1,659 to	Assistants (VHA)	–
		1,944	Robotics	367.5 million Hrs.
			Wearables	336.1 million Hrs.

Table 18.5 shows that the implementation of AI in the health and medical services sector saves enormous amounts of money in billions of dollars annually and saves lives in hundreds of thousands. Considering the potential economic impacts of AI in healthcare, a snapshot

economic analysis of AI in healthcare will also consider Exscientia's breakthrough in producing the first AI-designed drug (DSP-1181) to reach clinical trial as part of the revolutions accomplished by AI. This economic analysis will be based on our analytical framework.

18.10.2.1 ECONOMIC ANALYSIS OF THE IMPLEMENTATION OF AI IN THE HEALTH SECTOR

1. That the critical drivers of advancements in productivity are factors such as diffusion, innovation, and adoption of technology. Productivity in the healthcare sector encompasses several factors, such as quality of healthcare service provided, time to provide such services, and the effectiveness in terms of how the patient feels after receiving the service. Undoubtedly, AI positively impacts all the dimensions of productivity in the healthcare sector. One of the areas where AI's productive impact is being felt vividly is in the area of surgery. Most surgeries typically last long and require a long time for the patient to heal, often with devastating complications. However, AI-enabled non-invasive surgery with AR is a game-changer. According to Ref. [32], "Research shows that AI-assisted robotic procedures could result in almost five times lesser complications than surgeries operated traditionally." In addition, drug manufacturing is another area. The typical path to drug development takes about four to five years. However, Exscientia, in conjunction with the Japanese-based pharmaceutical company Sumitomo Dainippon Pharma, has recently invented a drug molecule using AI algorithms that have entered clinical trials in just 12 months. This is a revolutionizing stroke of AI's impact on healthcare. "The project in its entirety was five times faster than the usual drug discovery processes," said [33].
2. Cost-effectiveness. We will use Exscientia's AI-enabled drug production as a case in point. A synopsis of the cost-effectiveness of this AI-based drug manufacturing technology is shown using Table 18.6.

TABLE 18.6 Artificial Intelligence is Reducing the Cost of Drug Production

	AI Drug Manufacturing	**Traditional Drug Manufacturing**
Initial cost of investment	Cost of AI algorithm very low compared to the traditional method.	Costs $2 billion
Maintenance/operational cost	Low	Very high
Net present value	High	Low

Source: Adapted from Ref. [34].

From Table 18.6, it is evident that traditional drug discovery and manufacturing is costly and takes longer than the AI-backed method of drug discovery and manufacturing. The AI-backed method also delivers faster value with lower operational costs than the traditional method.

3. The key drivers of advancements in productivity are factors such as diffusion, innovation, and adoption of technology. The most relevant technology of this era is an IT. The labor landscape in healthcare is experiencing its fair share of transformations in the face of digitization and automation hitting the industry. AI will not replace humans in the workplace but would instead enhance the ability of healthcare professionals to deliver quality patient care and the ability to cope with the surging demands for medical care. Many traditional processes are already being digitized or automated, giving rise to a labor landscape similar to what we saw in the Exscientia and Sumitomo Danippon Pharma: a situation where healthcare experts partner with Ai systems providers to scale innovation. This may well be another reality of labor in healthcare as AI continues to transform it. According to Ref. [35], "Contract management and contract development organizations are, in recent years, taking over the more traditional roles of management and control of the means of production usually played by larger pharmaceutical companies. This has become obvious as renowned pharmaceutical companies like AstraZeneca, GSK, Pfizer, and others are increasingly outsourcing the production of their branded drugs." This new option will further reduce costs as companies may not bother with the massive cost of acquiring equipment but will only pay a flat fee for using efficient equipment akin to paying for cloud services.

4. Income inequality can be added in different countries due to latest computerized technologies. This is not readily the case in the instances we have used in this chapter. However, we argue that technology is a significant equipment for work, while the disparity in levels of education is the primary determinant of income inequality.

18.10.2.2 EXPERIMENTAL ANALYSIS OF TECHNOLOGY IMPLEMENTATION IN HEALTHCARE

This section will look at various projections of how the implementation of technology will affect and possibly transform healthcare. We will likely continue witnessing the increased implementation of digitized, automated, and efficient workflows. For instance, the process of seeing a doctor for patients who prefer telemedicine will most likely involve paying with the hospital app on their mobiles. For verification, the generated receipt can be sent to an online assistant. Before being linked to a doctor for consultation, the patient can enter the vital signs from the wearable devices.

Whereas patients who prefer face-to-face consultation will likely pay through their mobile apps or to a humanoid teller, an AI Nurse Assistant will verify their payments and obtain their vital signs before passing them on to the Doctor. This will achieve all the pre-consultation activities in a few minutes and increase efficiency and patient satisfaction. In the consulting room, the Doctor asks the patient to explain their symptoms, and a device like Watson takes the patient's history and gives a prescription. Then, the Doctor edits and certifies the prescription. This will also reduce consultation time and increase effectiveness. Then, in the Lab, AI-enabled machines will have the capacity to turn out tests that generally run for weeks in a couple of minutes. The same will happen in the pharmacy where AI-powered robots will receive prescriptions, sort them out from the shelves, and bring them to the pharmacist, checking, and approving the drugs and advising patients on using them.

Technology implementation in healthcare will most likely change the labor landscape in two ways. Firstly, we may see a scenario where data scientists and machine learning experts will become critical makeup of the healthcare workforce to help Doctors make sense of data and achieve better diagnosis in areas that have been fraught with misdiagnosis. However, this scenario will likely face challenges, given that data scientists and machine learning experts did not take the Hippocratic oath and, therefore, are not qualified to have access to patients' private health information and use them to make diagnoses. In addition, this scenario is subject to abuse as these experts could start giving unauthorized prescriptions using algorithms.

The second scenario of technology implementation in healthcare will involve humanoids as health workers. With humanoids like Sophia being given citizenship [36], they can equally be employed as first responders in the infectious diseases department, especially during epidemics and pandemics, to reduce deaths through infections. In this case, doctors and robots will walk as a team in attending to patients, as is already the case in robotic surgery. Next, the most crucial factor that defines and will continue to define how technology evolves in healthcare and the entire service industry is Data. Data is the new oil of technology. Machines can only be as intelligent as the data they are trained with. The proliferation of software as a medical service (SAMD) is concerning and will only retard the rate of adoption of this software.

This is traceable in part to the differences in standards used by different authorization bodies like FDA and others around the globe. How would you convince Asians, for instance, to accept software as a medical service technology that was trained with data from the U.S. alone? Evolving a regulatory system for SAMD akin to Bitcoin or cryptocurrency mining will address many inadequacies of the system and deliver many benefits. This system will ensure that diversity of Data is implemented. As a result, there is a need to have a system for the reliability of AI technology.

18.10.3 ENTERTAINMENT SECTOR

The entertainment industry is as old as man and has shown to be a significant attraction to man, a significant source of fun and relaxation, and an avenue for expressing humanity's unending creativity. Over time, the way and manner entertainment are presented have been evolving. However, the human element in entertainment has remained constant and may well remain so, even in this era of sweeping AI transformations.

It is important to note, at this stage, that what has set the entertainment industry apart is the originality of human creativity, which in itself is limitless. Importantly, we will limit our definition of entertainment to those art forms that are presented by humans with the sole aim of entertainment and serving fun, like music, singing, acting, and comedy. This scoping is important because some people include sports in the entertainment list. Having clarified that, it is important to consider how technology has impacted this sector.

Artists have used different technologies to support their creativity from the earliest times. Musicians have been using and still use tools like gongs, flutes, harps, and xylophones to support the melodies of their music. Now, other technologies like drums, electric drums, guitars, keyboards, trumpets, and others have been added. It is important to note that it is in the media element of entertainment that technology plays a role. In contrast, the human element always almost remains constant, except in cases of imitation. Like in other sectors, AI is transforming the way entertainment is served (that is, the media element) but can only imitate the expressed form of the human creativity existing in the form of data. It is critical to distinguish human creativity in terms of entertainment is infinite and can only be imitated. Similarly, Sandra [37] confirms that AI's influence is felt mainly in the media element of entertainment when she said: The application of AI in the media and entertainment sector can be categorized into four as follows:

1. **Marketing and Advertising:** This involves the deployment of AI applications in developing and serving marketing programs and advertisements for organizations and businesses with advanced capabilities for increased efficiency and effectiveness.
2. **User Experience Personalization:** This involves the utilization of AI applications in making business sense of users' data to enhance the provision of customized and personalized services to users that will meet their individual preferences and behavior.
3. **Search Optimization:** This involves using AI applications to optimize media and entertainment content for search functionalities. It will boost customer satisfaction and the efficiency of the overall search process.
4. **Innovative Experience with Broader Usage of VR and AR:** It will involve an advanced application of AI enhancing the way users experience media content using virtual and AR.

For instance, Morgan, a film trailer developed by IBM, the AI creative director code-named Ai-CD B by McCann Erickson Japan. Who was employed to work alongside other 11 employees and the enterprise visual intelligence (EVI) platform by Zorro Corporation designed to enable users to search for visuals on databases and perform analytics? But machines can only imitate this to a certain degree [38].

18.10.3.1 ECONOMIC ANALYSIS OF THE IMPLEMENTATION OF TECHNOLOGY IN THE ENTERTAINMENT INDUSTRY

In this section, as in other sections, we will apply our analytical framework to the entertainment industry to ascertain the effects of the implementation of technology in the industry:

1. The key drivers of advancements in productivity are factors such as diffusion, innovation, and adoption of technology. The most relevant technology of this era is an IT. AI is already transforming the media element of entertainment with innovative technologies that increase efficiency. These technologies are mostly applied to movie production and marketing, content categorization and personalization, and video games. For instance, the film trailer, Morgan, was developed within 24 hours using AI applications, whereas the traditional process lasted for weeks. Furthermore, with the development of AI platforms like EVI, it is hoped that AI-enabled platforms will increase the efficiency of entertainment production and marketing processes and contribute to cost reduction. This is most applicable in targeting advertisements more effectively to customers who need them and are most likely to pay for them. This strategy is already working with Netflix and will receive the wider application as AI continues its transformative journey [38].
2. Cost-effectiveness. In the case study, Morgan, the estimated budget was 8 million dollars, whereas the movie scarcely grossed approximately $8 million in sales globally. This confirms our claims that humans will always prefer and pay more for human creativity than machine-generated entertainment. It is also an indication that human-produced entertainment with technological support will be more cost-effective than artificially produced entertainment.

3. That technology has revolutionized employment by initiating machine-driven task-performance capabilities, which can continue to do so. AI is already transforming the labor landscape of the entertainment sector, as seen in the case of the creative director above, code-named AI-CD B hired by McCann Erickson Japan on April 01, 2016, alongside 11 human employees. We have reiterated the unlikeliness of robots or humanoids replacing human talent in the entertainment industry. This is more because humans have emotional and psychological attachments to other humans that machines can hardly replace. Humanoids may act in some movies, but this may only be regarded as fiction, which is not new to humans, but an extension of man's capacity to project beyond the ordinary.
4. That latest computerized technologies seem to have added to the rising income inequality and may likely continue provided they keep replacing skills and tasks hitherto identified with low-wage, menial, or middle-wage occupations. This is the same outcome as in the previous cases. Education rather than technology is the sole initiator and determinant of income inequalities.
5. Experimental analysis. Having mentioned the first experiment of hiring a humanoid alongside human employees, it may become a common experiment to see more humanoids engaged in certain roles in the entertainment industry. This is most likely going to involve the use of entertainment instruments with a finite set of keys like keyboards, Organs, and others. Humanoids could be trained with datasets from the best keyboardists like Herbie Hancock.

18.11 DIVERSITY AS A SERVICE (DAAS)

Suppose any service industry sector yearns for urgent implementation of technology for effective and efficient service delivery and more strategic practice. In that case, it is the Diversity and Inclusion sector. Technology is the most significant enabler of globalization. Globalization is the most excellent driver of diversities – creating opportunities, virtual, and physical, for people from different cultural, religious, racial, geographical, and other backgrounds to interact.

Historically, some of these interactions have led to some of humanity's most bitter and lingering problems. Hardly will a day pass without news of diversity-related issues of alarming proportions hitting the airwaves. Paradoxically, technology is used to transport these unfortunate incidents across the world in a matter of minutes. However, the same technology is the elixir to these problems [39, 40]. We recommend in this section the implementation of DaaS as the fourth component of the cloud infrastructure. The metaverse adventure accentuates this need since the platform will contain devices that collect people' expressions, body language, and voice quality as data resources for interpersonal interactions and relationships. Therefore, DaaS is the aspect of cloud computing that will be charged with gathering, recording, storing, and processing data that will help combat challenges to healthy human interactions and improve human relations in and out of the workplace nationally and globally. Data is the oil of today's technology, and there is humungous data on Diversity and inclusion that could be processed for the good of all. We attempt a breakdown of DaaS as follows:

1. **Cross-Cultural Services:** This will encompass all the data on cross-cultural intelligence (CQ), pooling resources from global cultures and their interactions.
2. **Diversity Training/Education:** This will include all the data on the bias, business case for diversity, diversity training, Employee Resource Groups, etc.
3. **Gender Equality:** This includes all the information and data showing where women are in the socio-economic totem pole because of their gender.
4. **Human Resources:** This includes all the data that will show the human resources data of organizations and how they square in terms of inclusion.
5. **Gamification Services:** This will embody an app for creating various games for diverse groups and communities. For instance, a game could be created to model fruitful and cordial relationships between black communities and the police in the U.S., or a game to stimulate healthy interactions between neurotypicals and neurodivergent employees in the workplace, etc.

6. **Listening Services:** This service will include all the resources that will enable diversity professionals to listen to all the conversations on race, religious intolerance, bias, inequalities, discriminations, prejudice, stereotypes, etc., and enable them to develop contents that will dispel a breakdown in healthy interactions.

We hope that DaaS will culminate in Diversity Analytics and increase productivity in global diversity practice when fully developed.

18.12 CONCLUSION

From the preceding, it is evident that technology implementation has been transforming the service industry and will continue to transform it. The consequent economic implications of its implementation, like increased productivity, enhanced customer satisfaction, enablement of more innovative work options, are the driving forces of its adoption in the industry. Even though AI is the latest technology driving innovations in the service industry, it is doing so alongside chatbots, AR, VR, search, cloud computing, mobile, social, and IoT. In addition, all aspects of human endeavors will likely witness the pervasive influence of machines. However, their effectiveness and utility will only depend on the level of human acceptance and adoption.

Technology is innovating cost efficiency, productivity, and replacing strenuous inefficient jobs with healthier and smarter options in the logistics sector. In the health sector, AI-backed processes save vast amounts of money in billions of dollars and human lives in hundreds of thousands.

However, no matter how powerful and valuable machines evolve, humans will always feel more at home and enjoy human entertainment more than machine-generated entertainment. They will most likely treat the way they treat fiction. Finally, we have seen from the preceding that we are now in a knowledge economy, where the minimum requirement for trade is the possession of knowledge enough to turn your mobile phone into a source of income. Therefore, technology is not the chief enabler of income inequities but disparities in levels of education. However, technology has shown to be the leveler, given its role in creating opportunities for people who could not have formal education to acquire it through Massive Open Online Courses. These online platforms offer financial aid to the downtrodden in society and grant them access to some of the best institutions of learning in the world. Therefore, the promises of implementing technology in the service sector outweigh its flipsides.

KEYWORDS

- **Amazon robotics**
- **artificial intelligence**
- **cloud computing**
- **diversity as a service**
- **economic analysis**
- **healthcare sector**
- **service industry**

REFERENCES

1. Mass. F., (2020). *IDC Survey Finds Artificial Intelligence Adoption Being Driven by Improved Customer Experience, Greater Employee Efficiency, and Accelerated Innovation.* https://www.businesswire.com/news/home/20200610005127/en/IDC-Survey-Finds-Artificial-Intelligence-Adoption-Being-Driven-by-Improved-Customer-Experience-Greater-Employee-Efficiency-and-Accelerated-Innovation (accessed on 07 September 2023).
2. Fortune Business Insights, (2021). *Artificial Intelligence (AI) Market Size, Share & Covid-19 Impact Analysis, By Component (Hardware, Software, and Services), By Technology (Computer Vision, Machine Learning, Natural Language Processing, and Others), By Deployment (Cloud, On-premises) By Industry (Healthcare, Retail, IT & Telecom, BFSI, Automotive, Advertising & Media, Manufacturing, and Others), and Regional Forecast, 2021 – 2028.* Available: https://www.fortunebusinessinsights.com/industry-reports/artificial-intelligence-market-100114 (accessed on 07 September 2023).
3. Hank, S., (2017). *Eight Ways Technology is Changing the Service Industry.* https://www.infoworld.com/article/3227548/eight-ways-technology-is-changing-the-service-industry.html (accessed on 07 September 2023).
4. Bill, W., (2017). *JD.com: Building the Smart Logistics System for the Future.* https://d3.harvard.edu/platform-rctom/submission/jd-com-building-the-smart-logistic-system-for-the-future/ (accessed on 07 September 2023).

5. Ornaith, K., (2013). *Four Reasons Why Site Search is Important for Online Retailers.* Available: https://econsultancy.com/four-reasons-why-site-search-is-vital-for-online-retailers/ (accessed on 07 September 2023).
6. Dave, C., (2020). *Search Engine Marketing Statistics 2020.* Available: https://www.smartinsights.com/search=engine-marketing/search-engine-statistics/ (accessed on 07 September 2023).
7. Taehyee, U., & Kim, T., (2020). How does an intelligent chatbot Affect Customers compared with self-service Technology for sustainable services. *Sustainability. 12*(12), 5119. doi: 10.3390/su12125119/.
8. Enisa, K., (2020). *Chatbots in Financial Services: Benefits, Use Cases and Key Features.* Available: https://www.artificial-solutions.com/blog/chatbots-financial-services-benefits-use-cases (accessed on 07 September 2023).
9. Insider Intelligence, (2021). *Chatbot Market in 2021: Stats, Trends, and Companies in the Growing AI Chatbot Industry.* Available: https://businessinsider.com/chatbot-market-stats-trends?IR (accessed on 07 September 2023).
10. Ayat, S., (2019). *Chatbots In Customer Service – Statistics and Trends [Infographic].* Available: https://www.invespcro.com/blog/chatbots-customer-service/ (accessed on 07 September 2023).
11. Yehia, E., (2020). *Augmented/ Virtual Reality: Market Size and Applications.* Available: https://infomineo.com/augmented_virtual_reality_market_size_applications/ (accessed on 07 September 2023).
12. Andrew, M., (2020). *Rise of M-Commerce: Mobile Ecommerce Shopping Stats & Trends in 2021. Available:* https://www.businessinsider.com/mobile-commerce-shopping-trends-stats?IR=T (accessed on 07 September 2023).
13. Naiya, S., (2019). *How IoT is Impacting Various Industries Worldwide.* Available: https://jaxenter.com/iot-industry-impact-160837.html (accessed on 07 September 2023).
14. FinanceOnline, (2021). *154 Impressive IoT Statistics: 2020/2021 Data Analysis & Market Share.* Available: https://financesonline.com/iot-statistics/ (accessed on 07 September 2023).
15. National Academies of Sciences, Engineering, and Medicine, (2017). *Information Technology and the U.S. Workforce: Where Are We and Where Do We Go from Here?* Washington, DC: The National Academies Press. doi: 10.17226/24649.
16. LogisticsIQ, (2019). *Warehouse Automation: Rise of Warehouse Robots.* Available: https://www.roboticsbusinessreview.com/wp-content/uploads/2019/10/RiseOfTheWarehouseRobots-LogisticsIQ.pdf (accessed on 07 September 2023).
17. Liu, Y., (2017). *How Emerging Technologies are Shaping China's Logistics Industry.* Available: http://chinaplus.cri.cn/news/china/9/20170417/3112.html (accessed on 07 September 2023).
18. Lili, (2017). *Is Alibaba's Smart Logistic Platform Sufficient?* Available: https://digital.hbs.edu/platform-rctom/submission/is-alibabas-smart-logistic-platform-sufficient (accessed on 07 September 2023).
19. Sean, G., (2020). *JD Logistics: Enabling Smart Supply Chain Solutions.* Available: https://supplychaindigital.com/logistics-1/cainiao-smart-logistics-announces-new-alibaba-strategy (accessed on 07 September 2023).

20. JD.com. (2021). *Our Business.* Available: https://corporate.jd.com/ourBusiness (accessed on 07 September 2023).
21. Jason, D. R., (2019). *How Robots Are Transforming Amazon Warehouse Jobs – for Better and Worse.* Available: https://www.vox.com/recode/2019/12/11/20982652/robots-amazon-warehouse-jobs-automation (accessed on 07 September 2023).
22. Charlie, C., (2020). *China's Cainiao is Revolutionizing How Goods Get Delivered. Will the Rest of the World Follow its Rules?* Available: https://ca.movies.yahoo.com/chinas-cainiao-revolutionizing-goods-delivered-021810664.html (accessed on 07 September 2023).
23. Craig, A., (2019). *3 Ways Robots Affect the Economy.* Available: https://www.investopedia.com/articles/markets-economy/091316/3-ways-robots-affect-economy.asp (accessed on 07 September 2023).
24. Alizilia, (2015). *Cainiao, Alibaba's Logistics Arm, Opens Up.* Available: https://www.alizila.com/cainiao-alibabas-logistics-arm-opens/ (accessed on 07 September 2023).
25. Sam, D., (2020). *32 Examples of AI in Healthcare That Will Make You Feel Better ABOUT The Future.* Available: https://builtin.com/artificial-intelligence/artificial-intelligence-healthcare (accessed on 07 September 2023).
26. Codrin, A., (2020). *Artificial Intelligence in Healthcare: The Future is Amazing.* Available: https://healthcareweekly.com/artificial-intelligence-in-healthcare/ (accessed on 07 September 2023).
27. Seo, S. W., & Yoon, H., (2020). Artificial intelligence in healthcare: Current applications and issues. *J Korean Med. Sci., 35*(48), e425. https://doi.org/10.3346/jkms.2020.35.e425.
28. Oleksii, T., (2021). *Technology Trends in Healthcare in 2021: The Rise of AI.* https://mobider.viz/blog/technology-trends-healthcare-digital-transformation (accessed on 07 September 2023).
29. Lansiti, M., & Lakhani, K. R., (2017). The truth about blockchain. *Harvard Business Review*. From the January-February 2017 issue. Available: https://hbr.org/2017/01/the-truth-about-blockchain (accessed on 07 September 2023).
30. Accenture, (2017). *Artificial Intelligence: Healthcare's New Nervous System.* Available: https://www.accenture.com/t2017-121570320592-w-/us-en/-acamedia/PDF-49/Accenture-Health-Artificial-Intelligence.pdf (accessed on 07 September 2023).
31. Buindo, E., et al., (2020). *The Socioeconomic Impact of AI in Healthcare.* Available: https://www.medtecheurope.org/wp-content/uploads/2020/10/mte-ai_impact-in-healthcare_oct2020_report.pdf (accessed on 07 September 2023).
32. Forbes, Bernard Marr, "How Is AI Used In Healthcare – 5 Powerful Real-World Examples That Show The Latest Advances," [Online available on: https://www.forbes.com/sites/bernardmarr/2018/07/27/how-is-ai-used-in-healthcare-5-powerful-real-world-examples-that-show-the-latest-advances/?sh=661289945dfb] (accessed on 07 September 2023).
33. Koh, D., (2020). *Sumitomo Dainippon Pharma and Exscientia Achieve Breakthrough in AI Drug Discovery.* Available: https://www.healthcareitnews.com/news/apac/sumitomo-dainippon-pharma-and-exscientia-achieve-breakthrough-ai-drug-discovery (accessed on 07 September 2023).

34. Paul, D., Sanap, G., Shenoy, S., Kalyane, D., Kalia, K., & Tekade, R. K., (2021). Artificial intelligence in drug discovery and development. *Drug Discovery Today, 26*(1), 80.
35. AiChe., (2018). *Pharmaceutical Manufacturing: Current Trends and What Next.* Available: https://www.aiche.org/resources/publications/cep/2018/december/pharmaceutical-manufacturing-current-trends-and-whats-next (accessed on 07 September 2023).
36. Paviainen, J., & Coeckelbergh, M., (2020). *The Political Choreography of the Sophia Robot: Beyond Robot Rights and Citizenship to Political Performances for the Social Robotics Market.* Available: https://link.springer.com/article/10.1007/s00146-020-01104-w (accessed on 07 September 2023).
37. Sandra, P., (2020). *How AI is Transforming the Entertainment Industry.* Available: https://readwrite.com/2020/01/10/how-ai-is-transforming-the-entertainment-industry/ (accessed on 07 September 2023).
38. Kumbar, S., (2019). *AI in Movies, Entertainment, and Visual Media – 5 Current Use-Cases.* Available: https://emerj.com/ai-sector-overviews/ai-in-movies-entertainment-visual-media/ (accessed on 07 September 2023).
39. Das, S., (2021). Flattening the diversified sphere through digital inclusivity. In: Sen, K., & Shahi, S., (eds.), *Creating a Culture of Diversity and Inclusiveness in India Inc.: Responsible Leadership and Sustainable Management.* Springer, Singapore. https://doi.org/10.1007/978-981-16-4237-1_8.
40. Eze, I. E., Shukla, V. K., & Verma, A., (2021). Diversity and inclusion through technological trends. In: Sen, K., & Shahi, S., (eds.), *Creating a Culture of Diversity and Inclusiveness in India Inc.: Responsible Leadership and Sustainable Management.* Springer, Singapore. https://doi.org/10.1007/978-981-16-4237-1_11.

Index

H

J

K

L

M

N

O

P

Q

R